CZECHOSLOVAK ACADEMY OF SCIENCES

Aphid Parasites *(Hymenoptera, Aphidiidae)* of the Mediterranean Area

CZECHOSLOVAK ACADEMY OF SCIENCES

Scientific Editor: Doc. Dr. L. Weismann
Scientific Adviser: Prof. Dr. A. Pfeffer

Aphid Parasites
(Hymenoptera, Aphidiidae)
of the Mediterranean Area

PETR STARÝ

Springer-Science+Business Media, B.V.

Distribution throughout the world with the exception of Socialist countries:

ISBN 978-90-6193-029-7   ISBN 978-94-010-1561-5 (eBook)
DOI 10.1007/ 978-94-010-1561-5

CONTENTS

6

INTRODUCTION

Aphids are one of the insect groups whose economic importance increases with the development of agriculture. The extensive monoculture and some agrotechnical practices result in much better conditions for population increase than the aphids find in natural environments. Pesticide treatments against various pests have been followed by the occurrence of secondary pests and the aphids are agreed as belonging mainly to this group. The intensification of pesticide treatments has shown that the aphids cannot be eradicated but, on the contrary, resistant populations have appeared which exhibit even more vitality than the original sensitive strains. Besides that, the routine pesticide treatments have adversely influenced the parasite populations. Consequently many parasite species suffer both from the disappearance or significant reduction of their reservoirs due to extensive monoculture system and non-selective treatments in the cultivated areas; it is one of the aims of the research to concentrate upon such species that are able to occur perennially and have their reservoirs in the crop monocultures.

Biological control of pests in the broad sense is a necessary part of the integrated plant protection. In the classical scheme of the integrated control of pests, however, it is still dealt with in its primary importance, i. e., the effectiveness of either indigenous or introduced parasites is supported by other methods. As it is known, this primary importance of biological control has been neglected for years and integrated control seems to be a result of checking our approach to pest management. Then, from this point of view, the occurrence of resistant strains of aphids demonstrates and stresses the possible — if we may use this term — secondary importance of biological control. The latter can work in cases where the effectiveness of the pesticides has decreased to negligible value and their use has even resulted in very undesirable side effects such as resistant pest populations.

The aphidiid parasites of aphids are one of the groups whose utilization in biological control has given significant results in many countries of the world. The research on aphid parasites has also been included in international programmes.

The enormous increase in research on biological control after World War II has reflected in the corresponding amount of papers; moreover, biological control is a synthetic trend as the applied phase of any biological control programme must be based on the results of studies on the taxonomy, distribution, host range, etc., of the target species. In this respect the development and intensification of international working parties is very promising. However, the character of contemporary research requires good reviewing papers that bring the research on a certain group on a checked and useful basis and thus eliminate waste of time and inevitable inaccuracy connected with search and study of papers scattered in the world literature. A very good idea in this sense has been the publishing of the "Index of Entomophagous Insects" by the O. I. B. C. The elaboration of the aphidiids in this Index (cf. MACKAUER and STARÝ 1967) makes the selection of records pertaining to a certain area difficult both as to the fauna and problem of biological control, but it fulfills quite well its primary role — the critical elaboration of the world state in research on the group. It had been one of the points that stimulated the author to a summarizing elaboration of the aphidiid parasites of at least some of the zoogeographical areas of the world.

The Mediterranean area of the Palearctic which is the subject of the present paper is one of the very important areas of distribution of the aphidiid parasites for several reasons. It apparently represents one of the centres of origin of the group where today the more or less natural communities and extensive areas of the past, present and future cultivation can be found. The fauna of both aphids and parasites is very rich in this area and it has been used as a well-known source of parasite species for exportation to other countries. The climatic conditions that allow the occurrence of several generations of pests and parasites in a year are an indication of the suitability of this area for biological control activities.

The present paper is a critical synthesis and it deals with the characteristics and peculiarities of the Mediterranean as shown in the aphidiid parasites and their utilization in biological control of aphids, and it has been intended as a base for future research. It is based on the knowledge of the world fauna and biology of the group which have been elaborated in other summarizing papers (MACKAUER and STARÝ 1967, STARÝ 1970). The dead point for the records included in this paper is December 1974, with the exception of some papers by the author that have been used as manuscripts (in press) and published in the course of 1975.

I. REVIEW OF GENERA AND SPECIES

The scheme of the records pertaining to the particular species is as follows: (1) Synonymy (Syn.) includes only the names of the synonymized species; the corresponding references can be found in the taxonomic literature (for example, MACKAUER and STARÝ 1967, MACKAUER 1968). (2) Hosts (Hosts): this is a review of the host records in the Mediterranean; the particular countries are followed by the authors' names and years of publication in brackets. (3) Distribution (Distr.) includes a list of the countries where a species has been established in the Mediterranean; the corresponding records (references) should be searched in the review of the hosts. (4) Biology (Biol.) includes information on biological data, effectiveness, utilization in control, the references being mentioned in brackets. These records pertain to the Mediterranean area whereas more comprehensive information on other areas might be obtained in the summarizing and other papers (see MACKAUER and STARÝ 1967, MACKAUER 1968, STARÝ 1970).

The species mentioned in brackets [ ] are apparently mis-identifications but this does not exclude the occurrence of such a species in the Mediterranean.

Genus: *Aphidius* Nees 1819

Syn.: *Incubus* Schrank 1802. *Theracmion* Holmgren 1872. *Euaphidius* Mackauer 1961.

A. absinthii Marshall 1896

Syn.: ?*Bracon melanocephalus* Nees 1811. ?*Aphidius* (*Aphidius*) *lutescens* Haliday 1834. ?*Aphidius* (*Aphidius*) *asteris* Haliday 1834. *Aphidius commodus* Gahan 1927. ?*Aphidius cardui* Marshall var. *artemisiae* Ivanov 1927.

Hosts: *Brachycaudus persicae* Passerini — Spain (Chalver 1973) (!). *Macrosiphoniella absinthii* Linné — Italy (Starý 1966, 1973). *M. artemisiae* Boyer de Fonscolombe — s. France (Starý et al. 1971, 1973, Starý 1973). Corse (Starý et al. 1973). Italy (Starý 1966, 1973). Turkey. Gruzia. *M. helichrysi* Remaudière — Corse (Starý et al. 1971, 1973). *M. millefolii* DeGeer — Corse (Starý et al. in press). Gruzia. *M. pulvera* Walker — s. France (Starý et al. 1973, Starý 1973). *M. sanborni* Gillette —

Italy (Starý 1966, 1973). *M. staegeri* Hille Ris Lambers — s. France (Starý et al. 1971, Starý 1973). *M. tapuskae* Hottes et Frison — Corse (Starý et al. 1971, Starý 1973). s. France (Starý et al. 1973). *M.* sp. — Crimea (Starý 1965). Turkey. Azerbaidjan (Starý 1965). Gruzia (Achvlediani 1964, Starý 1965). *Staticobium limonii* Mordwilko — Corse (Starý et al. in press). s. France (Starý et al. in press). *Without host data* — s. France. Italy (Quilis 1932). Spain (Chalver 1973).

Distribution: Spain, s. France, Corse, Italy, Crimea, Turkey, Transcaucasia (Gruzia, Azerbaidjan).

Biol.: Relation to ants in Italy (Starý 1966).

A. cingulatus Ruthe 1859

Syn.: *Aphidius gregarius* Marshall 1872. *Aphidius pterocommae* Ashmead 1889. *Aphidius lachni* Ashmead 1889. *Aphidius pterocommae* Marshall 1896.

Hosts: *Pterocomma pilosum* Buckton — Italy (Tremblay 1967). *P. populeum* Kaltenbach — s. France (Starý et al. 1971, Starý 1973). Italy (Starý 1966, 1973). Gruzia. *P.* sp. — Italy (Starý 1966, 1973). Gruzia.

Distr.: s. France, Italy, Gruzia.

Biol.: Relation to ants in Italy (Starý 1966).

A. colemani Viereck 1912

Syn.: *Aphidius platensis* Brèthes 1913. *Aphidius hübrichi* Brèthes 1913. *Aphidius porteri* Brèthes 1915. *Aphidius aphidiphilus* Benoit 1955. *Aphidius leroyi* Benoit 1955. *Aphidius transcaspicus* Telenga 1958.

Hosts: *Aphis punicae* Passerini — Iraq (Al — Azawí 1970). *A. zizyphi* Theobald — Iraq (Starý and Kaddou 1971). *A.* sp. — Syria (Starý 1975). Turkey (Liste d'Ident. 1971). *Hyalopterus pruni* Geoffroy — Algeria (Starý 1975). Tunisia — ?(Rolli 1974). Spain (Starý and Remaudière 1973). Italy (Starý 1964, 1966, 1973, Roberti 1969, Tremblay 1967). Sicily (Starý 1966). Egypt (Liste d'Ident. 1971). Lebanon (Biol. Contr. Inf. Bull. 1967). Israel (Starý 1966, 1970, 1975). Iraq (Al — Azawí 1970, Starý 1969, Starý and Kaddou 1971, Al — Rawy et al. 1969). Gruzia. *Melanaphis donacis* Passerini — s. France (Starý 1975, Sharma 1965). Italy (Starý 1964, 1966, 1973). Iraq (Starý and Kaddou 1971). *Rhopalosiphum nympheae* Linné — Lebanon (Biol. Contr. Inf. Bull. 1967). *Toxoptera aurantii* Boyer de Fonscolombe — ?Israel (Rosen 1964, 1966, 1967, 1969, Avidov and Harpaz 1969).

Distr.: Spain, s. France, Algeria, Tunisia, Italy, Sicily, Turkey, Egypt, Lebanon, Syria, Israel, Iraq, Gruzia.

Biol.: Oviposition, development, sex ratio in France (Sharma 1965). Relation to ants in Italy (Starý 1966), Iraq (Starý 1969). Predation of Chrysopids on the mummies in Iraq (Al — Rawy et al. 1969). Effectiveness in Iraq (Al — Rawy et al. 1969, Starý 1969, Al — Azawí 1970), in France (Sharma 1965). Integrated control on peach in Italy (Starý 1964, 1966), peach and apricot in Iraq (Al — Rawy et al. 1969, Al — Azawí 1970), vegetables in s. France (Biliotti and Sharma 1965).

10

A. eglanteriae Haliday 1934

Hosts: *Chaetosiphon chaetosiphon* Nevsky — s. France (Starý et al. 1971, Starý 1973). *Ch. alpestre* Hille Ris Lambers — s. France (Starý et al. 1971, Starý 1973). Distr.: s. France.

A. ervi Haliday 1834

Syn.: ?*Aphidius infirmus* Nees 1834. *Aphidius ulmi* Marshall 1896. *Aphidius medicaginis* Marshall 1898. *Aphidius fumipennis* Györfi 1958. *Aphidius ervi* Haliday ssp. *nigrescens* Mackauer 1962 (partim). *Aphidius caraganae* Starý 1963. *Aphidius mirotarsi* Starý 1963.
Hosts: *Acyrthosiphon bidentis* Eastop — Algeria. *A. lambersi* Leclant et Remaudière — s. France (Starý et al. 1973). *A. pisum* Harris — Morocco (Eady 1969). s. France (Starý et al. 1971, 1973). Italy (Starý 1966, 1973, Tremblay 1967). Sicily (Starý 1966). Corse (Starý et al. 1973, Starý et al. in press). Bulgaria (Pělov 1972, Grigorov 1972). Israel (Avidov and Harpaz 1969, Bodenheimer and Swirski 1957). Lebanon (Mackauer and Finlayson 1967). Iraq. *Aphis* sp. — s. France. *Macrosiphum euphorbiae* Thomas — Algeria, Corse (Starý et al. in press). *M. inexpectatum* Leclant — Corse (Starý et al. 1973). *Microlophium evansi* Theobald — Italy (Starý 1966, 1973, Tremblay 1967). *Myzus persicae* Sulzer — Algeria. Corse (Starý et al. 1971). Italy (Starý 1973). Gruzia. *Sitobion avenae* Fabricius — Corse (Starý et al. in press). Yugoslavia. Bulgaria. *S. fragariae* Walker — Corse (Starý et al. in press), s. France. *S.* sp. — s. France. *Without host data* — Morocco (Mimeur 1934). Spain (Starý 1973, Chalver 1973). Iraq (Al — Azawí 1970). Lebanon (Mackauer and Starý 1967, Mackauer and Finlayson 1967).
Distr.: Spain, Algeria, Morocco, s. France, Corse, Italy, Sicily, s. Yugoslavia, Bulgaria, Lebanon, Israel, Iraq. Gruzia.

A. funebris Mackauer 1961

Syn.: ?*Aphidius cardui* Marshall var. *cirsii* Ivanov 1925. ?*Aphidius eriophori* Mackauer 1967 (in Mackauer and Starý 1967).
Hosts: *Aphis fabae* Scopoli — Gruzia (Achvlediani 1964) (!). *A. farinosa* Gmelin — Gruzia (Achvlediani 1964) (!). *Uroleucon carthami* Hille Ris Lambers — Sicily (Starý 1966, 1973). *U. chondrillae* Nevsky — s. France (Starý et al. 1971, Starý 1973). Corse (Starý et al. 1971, Starý 1973). *U. cichorii* Koch — s. France (Starý et al. 1973, Starý 1973). Corse (Starý et al. in press). Italy (Starý 1966, 1973). Bulgaria (Starý 1962, 1973). Gruzia. *U. inulae* Ferrari — s. France (Starý et al. 1971, Starý 1973). Corse (Starý et al. 1971, Starý 1973). *U. jaceae* Linné — s. France (Starý et al. 1973). Italy (Starý 1966, 1973, Roberti 1969). Sicily (Starý 1966, 1973). Bulgaria (Starý 1962, 1973). Iraq (Starý and Kaddou 1971). Gruzia (Starý 1965). *U. ochropus* Hille Ris Lambers — s. France (Starý et al. 1973, Starý 1973), *U. picridis* Fabricius — Corse (Starý et al. 1971, Starý 1973). *U. sonchi* Linné — s. France (Starý et al. 1971, Starý 1973). Corse (Starý et al. in press). Italy (Starý 1973). Sicily (Starý 1966, 1973, Roberti **11**

1969). Bulgaria (Starý 1962). Iraq (Starý and Kaddou 1971, Al — Azawí 1970). *U.* sp. — Corse (Starý et al. in press). Italy (Starý 1973). Iraq (Starý 1966). Gruzia. *Without host data* — Algeria (Starý 1962). s. France. Italy (Mackauer and Starý 1967). Iraq (Liste d'Ident. 1971).

Distr.: Algeria, s. France, Corse, Italy, Sicily, Bulgaria, Iraq, Gruzia.

A. hieraciorum Starý 1962

Hosts: *Nasonovia nigra* Hille Ris Lambers — s. France (Starý al. 1973). Corse (Starý et al. 1971). Spain (Starý and Remaudière 1973). *N. ribisnigri* Mosley — s. France (Starý et al. 1973).

Distr.: Spain, s. France, Corse.

A. hortensis Marshall 1896

Syn.: *Aphidius (Aphidius) berberidis* Smith 1944.

Hosts: *Liosomaphis berberidis* Kaltenbach — Italy (Starý 1966). Gruzia (Starý 1965). *Without host data* — Italy (Mackauer et Starý 1967). Gruzia.

Distr.: Italy, Gruzia.

A. matricariae Haliday 1834

Syn.: *Aphidius (Aphidius) cirsii* Haliday 1834. *Aphidius (Aphidius) arundinis* Haliday 1834. *Aphidius phorodontis* Ashmead 1889. *Aphidius chrysanthemi* Marshall 1896. *Aphidius polygoni* Marshall 1896. *Aphidius lychnidis* Marshall 1896. *Aphidius valentinus* Quilis 1931. *Aphidius affinis* Quilis 1931. *Aphidius arundinis* Haliday var. *obscuriforme* Quilis 1931. *Aphidius discrytus* Quilis 1931. *Aphidius merceti* Quilis 1931. *Aphidius baudyši* Quilis 1931. *Aphidius renominatus* Hincks 1943. *Aphidius nigriteleus* Smith 1944.

Hosts: *Acyrthosiphon lambersi* Leclant and Remaudière — s. France (Starý et al. 1973). *Aphis affinis* Del Guercio — Iraq (Starý and Kaddou 1971). *A. craccivora* Koch — Corse (Starý et al. 1973). Iraq (Starý and Kaddou 1971). *A. fabae* Scopoli — Corse (Starý et al. in press). Italy (Tremblay 1967). Cyprus (Willkinson 1926). *A. frangulae* Kaltenbach — s. France (Starý et al. 1973). Corse (Starý et al. in press). *A. gossypii* Glover — Italy (Roberti 1969). Corse (Starý et al. in press). *A. hederae* Kaltenbach — Corse (Starý et al. in press). *A. lambersi* Börner — Corse (Starý et al. in press). *A. nasturii* Kaltenbach — Algeria. *A. parietariae* Theobald — s. France (Starý et al. 1971, 1973). Italy (Tremblay 1967). Iraq (Starý and Kaddou 1971). *A. ruborum* Börner — Corse (Starý et al. 1973). *A. solanella* Theobald — Spain (Quilis 1931). *A. spiraecola* Patch — Corse (Starý et al. 1973). s. France. *A. umbrella* Börner — Corse (Starý et al. in press). Greece. *A.* sp. — Algeria. s. France (Starý et al. 1971). Italy (Starý 1966, 1973, 1974). *Brachycaudus cardui* Linnaeus — Italy (Roberti 1969, Starý 1973, 1974). *B. helichrysi* Kaltenbach — Algeria. s. France (Starý et al. 1971, 1973). Corse (Starý et al. in press). Italy (Roberti 1969, Tremblay 1967). *B. persicae* Passerini — Corse (Starý et al. in press). *B.* sp. — Corse (Starý

12

et al. in press). *Capitophorus carduinus* Walker — Sicily (Barbagallo 1974). *C. eleagni* DelGuercio — s. France (Starý et al. 1971, Starý 1973). Sicily (Barbagallo 1974). Gruzia. *C. hippophaes* Walker — Turkey. *C. horni* Börner — s. France. *C. inulae* Passerini — Corse (Starý et al. 1973). Italy (Tremblay 1967). Sardegna. *C. sp.* — Italy (Starý 1966, 1973, 1974). Turkey. *Dysaphis plantaginea* Passerini — Italy (Tremblay 1967). *Eucarazzia elegans* Ferrari — s. France. *Forda* sp. — Spain (Quilis 1931) (?). *Galiobium langei* Börner — Italy (Starý 1966, 1973, 1974). *Hayhurstia atriplicis* Linné — Corse (Starý et al. in press). *Myzus cerasi* Fabricius — Corse (Starý et al. in press). *M. cerasi veronicae* Walker — s. France (Starý et al. 1971, Starý 1973). *M. ornatus* Laing — s. France. *M. persicae* Sulzer — Algeria. Portugal. Corse (Starý et al. 1973). s. France (Starý et al. 1973, Lyon 1973). Italy (Starý 1966, 1973, 1974, Tremblay 1973, Tremblay and Iaccarino 1971, Roberti 1969). Israel (Avidov and Harpaz 1969, Avidov and Kotter 1966, Rosen 1964, 1966, 1967, 1969). Bulgaria (Starý 1974). Iraq (Starý and Kaddou 1971). *M. varians* Davidson — Corse (Starý et al. 1971, Starý 1973). *Phorodon humuli* Schrank — Corse (Starý et al. in press). s. France. *Rhopalosiphum maidis* Fitch — Gruzia. *R. padi* Linné — Portugal. *Schizaphis graminum* Rondani — s. France (Starý et al. 1971). *Toxoptera aurantii* Boyer de Fonscolombe — s. France. Corse (Starý et al. in press). Spain. Italy (Roberti 1969, Tremblay 1967). *Uroleucon inulae* Ferrari — Italy (Roberti 1969) (?). Spain (Quilis 1931) (?). *Without host data* — Spain (Chalver 1973). France (Shands et al. 1965). Italy (Mackauer and Starý 1967). N. Africa (Schlinger and Mackauer 1963). Israel (Schlinger and Mackauer 1963). Gruzia.

Distr.: Algeria, Spain, Portugal, s. France, Corse, Sardegna, Italy, Sicily, Bulgaria, Greece, Cyprus, Turkey, Israel, Iraq, Gruzia.

Biol.: Biological control in greenhouses in Italy (Tremblay 1973).

A. phalangomyzi Starý 1963

Hosts: *Macrosiphoniella oblonga* Mordwilko — s. France (Starý et al. 1973).
Distr.: s. France.

A. picipes (Nees 1811)

Syn.: *Aphidius (Aphidius) avenae* Haliday 1834. *Aphidius crithmi* Marshall 1896. *Aphidius pascuorum* Marshall 1896. *Aphidius granarius* Marshall 1896. ?*Lysiphlebus hungaricus* Györfi 1958. *Aphidius caraganae* Starý 1963 (partim).

Hosts: *Acyrthosiphon pisum* Harris — s. France (Starý et al. 1973). *Aulacorthum solani* Kaltenbach — Spain (Quilis 1929). *Dysaphis plantaginea* Passerini — Italy (Principi et al. 1967, Liste d'Ident. 1971). *Macrosiphum doronicicola* Leclant — s. France (Starý et al. 1973). *M. euphorbiae* Thomas — Italy (Roberti 1969). *Myzus persicae* Sulzer — Italy (Starý 1966, 1967, Roberti 1969). Spain (Quilis 1929). Corse (Starý et al. in press). *M. sp.* — Spain (Quilis 1929). *Sitobion avenae* Fabricius — s. France (Starý et al. 1971, 1973). Italy (Roberti 1969, Starý 1966, Tremblay 1967). Crimea (Starý 1965). *Toxoptera aurantii* Boyer de Fonscolombe — Spain (Quilis

13

1929) (?). *Without host data* — Canary Isl. (Mackauer 1962). Italy (Mackauer and Starý 1967). Spain — (Gómez — Menor 1965, Chalver 1973).

Biol.: Seasonal history and effectiveness in Italy (Principi et al. 1967). Biology in Spain (?, Quilis 1929). Integrated control on apple in Italy (Principi et al. 1967).

Distr.: Canary Isl., Spain, s. France, Italy, Crimea.

A. rhopalosiphi De Stefani 1902

Hosts: *Rhopalosiphum* sp. — Italy (DeStefani 1902).
Distr.: Italy.
Note: This is an unclear species of *Aphidius* (see Starý 1973).

A. ribis Haliday 1834

Syn.: *Lysiphlebus ribaphidis* Ashmead 1889. *Aphidius scabiosae* Marshall 1896. *Aphidius ribis* Ashmead 1898.
Hosts: *Cryptomyzus ballotae* Hille Ris Lambers — s. France (Starý et al. 1971, Starý 1973). *C. galeopsidis* Kaltenbach — s. France (Starý et al. 1971, Starý 1973).
Distr.: s. France.

A. rosae Haliday 1834

Syn.: *Aphidius rosarum* Nees 1834. ?*Aphidius xanthostoma* Bouché 1834. ?*Aphidius protaeus* Wesmael 1835. *Aphidius cancellatus* Buckton 1876.
Hosts: *Macrosiphum rosae* Linné — Portugal, s. France (Starý et al. 1971, 1973). Corse (Starý et al. in press). Italy (Starý 1966, 1973, Roberti 1969, Tremblay 1967). Sicily (Starý 1973). Greece (Zouliamis 1968). Iraq (Starý and Kaddou 1971). Crimea (Starý 1962, 1965). Gruzia (Starý 1965, Achvlediani 1974). *Without host data* — Italy (Mackauer and Starý 1967). Yugoslavia (Fahringer 1924). Canary Isl. (Mackauer 1962).
Distr.: Canary Isl., Portugal, s. France, Corse, Italy, Sicily, s. Yugoslavia, Greece, Iraq, Crimea, Gruzia.

A. salicis Haliday 1834

Syn.: *Aphidius restrictus* Nees 1834. *Aphidius duodecimarticulatus* Ratzeburg 1852. *Aphidius dauci* Marshall 1896.
Hosts: *Cavariella aegopodii* Scopoli — Corse (Starý et al. in press). *C.* sp. — Corse (Starý et al. in press). Gruzia. *Hyadaphis foeniculi* Passerini — Italy (Tremblay 1967).
Distr.: Corse, Italy, Gruzia.

A. setiger (Mackauer 1961)

Syn.: ?*Aphidius cardui* Marshall var. *aceri* Ivanov 1925.
Hosts: *Periphyllus aceris* Linné — s. France (Mackauer 1961). *P. bulgaricus* Tashev — s. France (Starý et al. 1971, Starý 1973). *P. testudinaceus* Fernie — Italy

14

(Rondani 1974). *P.* sp. — s. France (Starý et al. 1971). Turkey. Gruzia (Starý 1965). *Without host data* — Italy (Rondani 1874, as *A. protaeus* Wesm.).

Distr.: s. France, Italy, Turkey, Gruzia.

A. sonchi Marshall 1896

Hosts: *Aphis durantae* Theobald — Egypt (Hassan 1957) (!). *Hyalopterus pruni* Geoffroy — Israel (Swirski 1951, Bodenheimer and Swirski 1957) (!). *Hyperomyzus lactucae* Linné — s. France (Starý et al. 1971, Starý 1973). *H. picridis* Börner — s. France (Starý et al. 1971). *H.* sp. — Sicily (Starý 1966, 1973). *Rhopalosiphum maidis* Fitch — Egypt (Hassan 1957) (!).

Distr.: s. France, Sicily, Egypt, Israel.

A. urticae Haliday 1834

Syn.: *Aphidius longulus* Marshall 1896. *Aphidius lonicerae* Marshall 1896. *Aphidius silenes* Marshall 1896. ?*Aphidius euphorbiae* Marshall 1896. ?*Aphidius goidanichi* Quilis 1932. *Aphidius ervi* Haliday ssp. *nigrescens* Mackauer 1962 (partim). *Aphidius rubi* Starý 1962. *Aphidius silvaticus* Starý 1962. *Aphidius aulacorthi* Starý 1963. *Aphidius caraganae* Starý 1963 (partim). *Aphidius mirotarsi* Starý 1963 (partim).

Hosts: *Acyrthosiphon euphorbiae* — Turkey. *A. loti* Theobald — s. France (Starý et al. 1971, 1973). *A. pisum* Harris — s. France (Starý et al. 1971). Corse (Starý et al. 1973). Sicily. Italy. Gruzia. *Macrosiphum doronicicola* Leclant — s. France (Starý et al. 1973). *Masonaphis* sp. — Gruzia. *Microlophium evansi* Theobald — s. France (Starý et al. 1971, Starý 1973, Tremblay 1967). Spain (Starý 1973). *Sitobion avenae* Fabricius — Corse (Starý et al. in press). Spain (Starý 1973). *S. fragariae* Walker — Corse (Starý et al. in press). *S.* sp. — Sicily (Starý 1973). *Without host data* — Corse (Starý et al. in press). Spain (Mackauer and Starý 1967).

Distr.: Spain, s. France, Corse, Italy, Sicily, Turkey, Gruzia.

A. uzbekistanicus Luzhetzki 1960

Syn.: ?*Aphidius beltrani* Quilis 1931, ?*Aphidius macropterus* Quilis 1931. ?*Aphidius granarius* Marshall var. *pailloti* Quilis 1931. ?*Aphidius indivisus* Quilis 1931. *Aphidius impressus* Mackauer 1965.

Hosts: *Forda* sp. — Spain (Quilis 1931) (?). *Melanaphis donacis* Passerini — s. France (Sharma 1965) (?). *Metopolophium dirhodum* Walker — Spain (Quilis 1931). *Rhopalosiphum maidis* Fitch — s. France. *Schizaphis fritzmuelleri* Leclant — s. France (Starý et al. 1973). *Sitobion avenae* Fabricius — Spain (Starý 1973). s. France (Starý et al. 1973). Corse (Starý et al. in press). Italy (Starý 1973, Roberti 1969 — as *A. pascuorum* Marsh.). Sicily. Yugoslavia. Crimea. *S. fragariae* Walker — Spain (Quilis 1931). s. France. Corse (Starý et al. in press). *S.* sp. — s. France. Corse (Starý et al. in press). Italy (Starý 1973). Sicily (Starý (1973). Bulgaria. *Without host data* — Spain (Starý 1973).

Biol.: Notes — s. France (Sharma 1965).

Distr.: Spain, s. France, Corse, Italy, Sicily, Bulgaria, Yugoslavia, Crimea. **15**

A. sp.

Hosts: *Acyrthosiphon lambersi* Leclant et Remaudière — Corse (Starý et al. 1973). s. France (Starý et al. 1971). *A. scariolae* Nevsky — s. France (Starý et al. 1971). Iraq (Al — Azawí 1970). *A. sp.* — Corse (Starý et al. in press). *Aphis fabae* Scopoli — Turkey (Liste d'Ident. 1971). *A. gossypii* Glover — Gruzia (Rekač and Dobrecova 1933). *A. sp.* — s. France. *Brachycaudus helichrysi* Kaltenbach — s. France (Starý et al. 1971). *B. sp.* — Gruzia (Achvlediani 1964). *Dysaphis plantaginea* Passerini — Algeria. *D. sp.* — s. France. *Macrosiphum sp.* — Iraq (Starý and Kaddou 1971). *M. euphorbiae* Thomas — Corse (Starý et al. 1971). *Macrosiphoniella millefolii* DeGeer — Corse (Starý et al. in press). *Melanaphis pyraria* Passerini — s. France. *Microlophium evansi* Theobald — Italy (Tremblay 1961). *Myzus persicae* Sulzer — Corse (Starý et al. 1973). Morocco. Italy (Menozzi 1932). *Periphyllus* sp. — Gruzia (Achvlediani 1964). *Rhopalosiphum maidis* Fitch — Israel (Avidov and Harpaz 1969). *Schizaphis fritzmuelleri* Leclant — s. France (Starý et al. 1971). *Sitobion avenae* Fabricius — Greece (Mentzelos et al. 1964). *S. fragariae* Walker — s. France. *Toxoptera aurantii* Boyer de Fonscolombe — Cyprus (Wood 1963). *Without host data* - Morocco. Iraq (Liste d'Ident. 1971). Israel (Bodenheimer and Swirski 1957). "*Siphoninus phillyreae* Hal." — Morocco (Mimeur 1946) (!, *Aleyrodidae*).

Genus: *Areopraon* Mackauer 1959

A. lepelleyi (Waterston 1926)

Hosts: *Eriosoma lanuginosum* Hartig — Italy (Starý 1966). *E. patchae* Börner et Blunck — Gruzia. *E. ulmi* Linné — Gruzia.
Distr.: Italy, Gruzia.
Biol.: Relation to ants in Italy (Starý 1966).

Genus: *Diaeretiella* Starý 1960

D. rapae (M'Intosh 1855)

Syn.: ?*Aphidius vulgaris* Bouché 1834. *Aphidius (Trionyx) rapae* Curtis 1860. *Misaphidus halticae* Rondani 1877. *Trioxys piceus* Cresson 1880. *Lipolexis chenopodiaphidis* Ashmead 1889. *Aphidius brassicae* Marshall 1896. *Diaeretus californicus* Baker 1909. *Lysiphlebus crawfordi* Rohwer 1909. *Diaeretus nipponensis* Viereck 1911. *Diaeretus (Aphidius) obsoletus* Kurdjumov 1913. *Diaeretus napus* Quilis 1931. *Diaeretus croaticus* Quilis 1934. *Diaeretus plesiorapae* Blanchard 1940. *Diaeretus aphidum* Mukerjee et Chatterjee 1950.
Hosts: *Aphis acanthi* Schrank — Spain (Quilis 1931, Starý 1961, 1974). *A. umbrella* Börner — Greece (Starý 1974). *Brachycaudus amygdalinus* Schouteden — Italy (Roberti 1969). Gruzia. *B. helichrysi* Kaltenbach — Italy (Roberti 1969). Starý 1974. *B. rumexicolens* Patch — s. France (Starý et al. 1971). *B. sp.* — Gruzia. *Brevicoryne*

16

brassicae Linné — Algeria. Morocco. Spain (Chalver 1973, Starý 1974). s. France (Starý et al. 1971, Lyon 1968). Corse (Starý et al. 1971). Cyprus (Starý 1974). Italy (Roberti 1969, Starý 1966, Tremblay 1965, 1967). Sicily (Starý 1966). Egypt (Hafez 1965). Lebanon (Talhouk 1961). Iraq (Al — Azawí 1970). Gruzia. *Diuraphis noxius* Mordwilko — Algeria. *Dysaphis tulipae* Boyer de Fonscolombe — Algeria. *Hayhurstia atriplicis* Linné — Algeria. Azerbaidjan (Starý 1965). Gruzia. *H. cadiva* Walker — s. France. *Macrosiphoniella sanborni* Gillette — Iraq (Al — Azawí 1970). *Melanaphis donacis* Passerini — s. France (Sharma 1965). *Myzus persicae* Sulzer — Algeria. s. France (Lyon 1968, 1973, Starý et al. 1971, 1973, Uščekov et al. 1972). Italy (Roberti 1969, Tremblay 1967, Starý 1974). Corse (Starý et al. in press). Israel (Starý 1967, Avidov and Harpaz 1969, Avidov and Kotter 1966). Gruzia (Achvlediani 1964, Starý 1965). *Rhopalosiphum maidis* Fitch — s. France. Morocco (Mimeur 1936). *Schizaphis longicaudata* Hille Ris Lambers — Sicily (Starý 1966, 1974). *S. graminum* Rondani — Israel (Bodenheimer and Swirski 1957). *Toxoptera aurantii* Boyer de Fonscolombe — Italy (Roberti 1969). *Uroleucon sp.* — Iraq (Starý and Kaddou 1971). *U. sonchi* Geoffroy — Iraq (Starý and Kaddou 1971). *Without host data* — Algeria (Starý 1961). Morocco (Mimeur 1934). Canary Isl. (Mackauer 1962). Spain (Quilis 1931, Starý 1961). France. Italy (Martelli 1911). Libya (Mackauer and Starý 1967). Egypt (Mackauer and Starý 1967). Iraq (Liste d'Ident. 1971). Gruzia. "*Phyllotreta nigra* E. H." (*Col. Chrysomelidae* — Italy) (Rondani 1877) (!).

Distr.: Spain, Canary Isl., Algeria, Morocco, Libya, s. France, Italy, Sicily, Greece, Cyprus, Egypt, Lebanon, Israel, Iraq, Gruzia, Azerbaidjan.

Biol.: Development (Tremblay 1966). Description of mummy, meconium, biol. notes in Egypt (Hafez 1965). Effects on the host (Tremblay 1965). Intraspecific competition (Tremblay 1966). Effectiveness in Iraq (Al — Azawí 1970). Biological control in France (Biliotti and Sharma 1965), in greenhouses, mass — rearing, storage (Lyon 1968).

Genus: *Diaeretus* Förster 1862

D. leucopterus (Haliday 1834)

Syn.: *Aphidius exspectatus* Gautier et Bonnamour 1936.
Hosts: *Eulachnus rileyi* Will. — Sicily (Tremblay 1970). *E. tuberculostemmata* Theobald — Israel. *E.* sp. — Italy (Starý 1966). *Without host data* — Israel (Mackauer 1962). Spain (Mackauer and Starý 1967).
Distr.: Spain, Italy, Sicily, Israel.

Genus: *Ephedrus* Haliday 1833

Syn.: *Elassus* Wesmael 1835.

E. cerasicola Starý 1962   **17**

Hosts: *Capitophorus inulae* Passerini — s. France (Starý et al. 1971). *Myzus persicae* Sulzer — Corse (Starý et al. 1973).
Distr.: s. France, Corse.

E. lacertosus (Haliday 1833)

Syn.: *Ephedrus muesebecki* Smith 1944.
Hosts: *Without host data* — Spain (Chalver 1973).
Distr.: Spain.

E. minor Stelfox 1941

Syn.: ?*Aphidius brevicornis* Nees 1834
Hosts: *Without host data* — Spain (Chalver 1973).
Distr.: Spain.

E. nacheri Quilis 1934

Hosts: *Hayhurstia atriplicis* Linné — Italy (Starý 1966, 1974). Gruzia. *H. cadiva* Walker — s. France.
Distr.: s. France, Italy, Gruzia.

E. niger Gautier, Bonnamour, Gaumont 1929

Syn.: ? *Alisia aphidivora* Rondani 1848. *Ephedrus (Ephedrus) campestris* Starý 1962.
Hosts: *Aphis sambuci* Linné — Italy (Rondani 1848) (!). *Macrosiphoniella artemisiae* Boyer de Fonscolombe — s. France (Starý et al. 1971). *M. helichrysi* Remaudière — Corse (Starý et al. 1973). *M. millefolii* DeGeer — Corse (Starý et al. 1971). *M. pulvera* Walker — s. France (Starý et al. 1973). *M. sanborni* Gillette — Italy (Starý 1966). *M.* sp. — Gruzia. Azerbaidjan (Starý 1965). *Uroleucon achilleae* Koch — s. France (Starý et al. 1973). *U. cichorii* Koch — Bulgaria (Starý 1962). *U. hypochoeridis* Hille Ris Lambers — s. France (Starý et al. 1973). *U. inulae* Ferrari — Corse (Starý et al. 1971). Sicily (Starý 1966). Monaco. *U. jaceae* Linné — Italy (Starý 1966). *U. sonchi* Geoffroy — Italy (Rondani 1874). Bulgaria (Starý 1962). *U.* sp. — Corse (Starý et al. 1973). s. France. Gruzia. *Without host data* — Italy (Rondani 1848).
Distr.: s. France, Monaco, Corse, Italy, Sicily, Bulgaria, Gruzia.

E. persicae Froggatt 1904

Syn.: *Ephedrus nevadensis* Baker 1909. *Ephedrus nitidus* Gahan 1917. *Ephedrus vidali* Quilis 1934. *Ephedrus interstitialis* Watanabe 1941. *Ephedrus pulchellus* Stelfox 1941. *Ephedrus impressus* Granger 1949. *Ephedrus (Ephedrus) holmani* Starý 1958. *Ephedrus (Ephedrus) palaestinensis* Mackauer 1959.
Hosts: *Aphis craccivora* Koch — s. France. Israel (Avidov and Harpaz 1969, Mackauer 1963). Iraq. *A. fabae* Scopoli — Italy (Tremblay 1967). Iraq (Al — Azawí 1970). *A. frangulae* Kaltenbach — Corse (Starý et al. in press). *A. gossypii* Glover —

18

Iraq (Al — Azawí 1970). *A. hederae* Kaltenbach — Spain. *A. lichtensteini* Leclant and Remaudière — Corse (Starý et al. 1973). *A. medicaginis* Koch — Israel (Mackauer 1963). *A. ruborum* Börner — Israel (Mackauer 1959). *A. umbrella* Börner — Israel (Mackauer 1963). Corse (Starý et al. in press). *A. sp.* — Syria. Israel (Mackauer 1963). *Brachycaudus amygdalinus* Schouteden — s. France (Starý et al. 1973). Italy (Roberti 1969). Libya, Israel (Avidov and Harpaz 1969, Bodenheimer and Swirski 1957). *B. cardui* Linné — Greece. *B. helichrysi* Kaltenbach — s. France (Starý et al. 1971). Corse (Starý et al. in press). Italy (Roberti 1969). *B. mimeuri* Remaudière — Corse (Starý et al. 1973). *B. persicae* Passerini — s. France (Starý et al. 1973). Corse (Starý et al. in press). *B. sp.* — Italy (Starý 1966). Gruzia. *Brachyunguis tamaricis* Lichtenstein — Algeria. s. France (Starý et al. 1973). *B. sp.* — s. France. Italy (Starý 1966). Sicily (Starý 1966). *Dysaphis crataegi* Kaltenbach — Lebanon (Talhouk 1961). Gruzia. *D. devecta* Walker — Gruzia. *D. plantaginea* Passerini — s. France (Starý et al. 1971). Italy (Roberti 1969, Tremblay 1967). *D. pyri* Boyer de Fonscolombe — s. France (Starý et al. 1973). Corse (Starý et al. in press). *D. reaumuri* Mordwilko — Gruzia. *D. sorbi* Kaltenbach — s. France (Starý et al. 1973). Italy (Starý 1966). *D. sp.* — Italy (Starý 1966). *Hyalopterus pruni* Geoffroy — Gruzia. *Melanaphis pyraria* Passerini — s. France (Starý et al. 1971, 1973). *Myzus cerasi* Fabricius — s. France (Starý et al. 1973). Italy (Starý 1966). Gruzia (Achvlediani 1964, Starý 1965). *M. persicae* Sulzer — s. France (Starý et al. 1971). Corse (Starý et al. in press). Gruzia. *Rhopalosiphum maidis* Fitch — Israel (Avidov and Harpaz 1969, Mackauer 1959). Lebanon (Biol. Contr. Inf. Bull. 1967). *R. padi* Linné — Israel (Mackauer 1959). *Roepkea marchali* Börner — Gruzia (Achvlediani 1964, Starý 1965). *Toxoptera aurantii* Boyer de Fonscolombe — Italy (Roberti 1969, Tremblay 1967). Cyprus (Starý 1967, Wood 1963). Israel (Avidov and Harpaz 1969, Mackauer 1963, Rosen 1964, 1966, 1967, 1969, Starý 1967). *Without host data* — Algeria. Canary Isl. (Mackauer 1962). Spain (Quilis 1934, Chalver 1973). Cyprus. Lebanon (Mackauer and Starý 1967). Greece (Liste d'Ident. 1971). Syria (Mackauer 1963). Israel. Italy (Triggiani 1973, on almond).

Distr.: Spain, Canary Isl., Algeria, Libya, s. France, Corse, Italy, Sicily, Greece, Cyprus, Syria, Lebanon, Israel, Iraq, Gruzia.

Biol.: Relation to ants in Italy (Starý 1966). Effectiveness in Italy (Roberti 1969), Iraq (Al — Azawí 1970).

E. plagiator (Nees 1811)

Syn.: *Aphidius parcicornis* Nees 1834. *Ephedrus japonicus* Ashmead 1906.

Hosts: *Acyrthosiphon pisum* Harris — Gruzia. *Anuraphis sp.* — Gruzia. *Aphis bupleuri* Börner — s. France (Starý et al. 1971). *A. craccae* Linné — Corse (Starý et al. in press). *A. fabae* Scopoli — Corse (Starý et al. in press). Italy (Starý 1966, Tremblay 1967). Gruzia. *A. lichtensteini* Leclant and Remaudière — Corse (Starý et al. 1971). *A. parietariae* Theobald — s. France (Starý et al. 1971). *A. pomi* DeGeer — Bulgaria (Pělov 1972). *A. spiraecola* Patch — Corse (Starý et al. 1971). *A. sp.* — Greece (Zouliamis 1968). Gruzia. *Brachycaudus helichrysi* Kaltenbach — Italy (Roberti **19**

1969). Bulgaria (Pělov 1972). *B. rumexicolens* Patch — Gruzia. *B. spireae* Passerini — Gruzia. *B.* sp. — Gruzia (Starý 1965). Azerbaidjan. *Brachyunguis* sp. — Gruzia (Achvlediani 1964). *Capitophorus inulae* Passerini — Italy (Tremblay 1967, Tremblay and Calvert 1971). *Corylobium avellanae* Schrank — Corse (Starý et al. in press). *Dysaphis crataegi* Kaltenbach — Gruzia. *D. devecta* Walker — Crimea (Starý 1962, 1965). Bulgaria (Pělov 1972). *D. plantaginea* Passerini — Bulgaria (Pělov 1972). *D. pyri* Boyer de Fonscolombe — Bulgaria (Pělov 1972). *Hyadaphis foeniculi* Boyer de Fonscolombe — Italy (Starý 1966). *Hyalopterus pruni* Geoffroy — Italy (Starý 1966). Bulgaria (Pělov 1972). Crimea (Starý 1965). Gruzia. *Liosomaphis berberidis* Kaltenbach — Gruzia. *Macrosiphum doronicicola* Leclant — s. France (Starý et al. 1973). *M. euphorbiae* Thomas — Corse (Starý et al. in press). *M. holmani* Leclant — Corse (Starý et al. 1971). *M. rosae* Linné — s. France (Starý et al. 1971). Gruzia. *M. saniculae* Leclant — Corse (Starý et al. in press). *M.* sp. — Turkey. *Myzus cerasi* Fabricius — Bulgaria (Pělov 1972). *M. varians* Davidson — Corse (Starý et al. 1971). *Rhopalosiphum insertum* Walker — Italy (Starý and Schlinger 1967, Vidano 1959). *Sitobion avenae* Fabricius — Corse (Starý et al. 1971). Italy (Starý 1966). Gruzia. *S. fragariae* Walker — Corse (Starý et al. in press). *Without host data* — Spain (Marshall 1896, 1899), Italy (Rondani 1848, Marshall 1899, Triggiani 1973, on almond). Iraq (Mackauer 1963). Gruzia (glasshouse).

Distr.: Spain, s. France, Corse, Italy, Bulgaria, Greece, Turkey, Iraq, Crimea. Gruzia.

Biol.: Italy (Vidano 1959).

E. sp.

Hosts: *Hyadaphis foeniculi* Passerini — s. France (Starý al. 1973), *Hyadaphis* sp. — s. France. *Myzus cerasi* Fabricius — Azerbaidjan (Starý 1965). *Phorodon humuli* Schrank — s. France. *Without host data* — s. France (Starý et al. 1971). Israel (Bodenheimer and Swirski 1957).

Genus: *Lipolexis* Förster 1862

Syn.: *Gynocryptus* Quilis 1931

L. gracilis Förster 1862

Syn.: *Gynocryptus pieltaini* Quilis 1931. *Aphidius palpator* Gautier et Bonnamour 1931.

Hosts: *Anoecia corni* Koch — Italy (Starý 1966). *Aphis arbuti* Ferrari — Corse (Starý et al. in press). *A. cisticola* Leclant et Remaudière — Corse (Starý et al. 1971). *A. confusa* Walker — s. France (Starý et al. 1971). *A. craccivora* Koch — Italy (Starý 1966, 1967). Corse (Starý et al. 1971). *A. dorycnii* Hille Ris Lambers — s. France (Starý et al. 1971). *A. fabae* Scopoli — s. France (Starý et al. 1971), Corse (Starý et al. in press), Italy (Starý 1966, Roberti 1969). Gruzia. *A. hederae* Kaltenbach — Italy

20

(Starý 1966). *A. pomi* DeGeer — Italy (Starý 1966). *A. ruborum* Börner — Italy (Starý 1966). Sicily (Starý 1966). *A. sedi* Kaltenbach — Italy (Starý 1966). *A. solanella* Theobald — Greece (Zouliamis 1968). *A. vallei* Stroyan et Hille Ris Lambers — Italy (Starý 1966). *A.* sp. — Algeria. s. France. Corse. Italy (Starý 1966). Sicily (Starý 1966). Monaco. Gruzia. *Brachycaudus amygdalinus* Schouteden — Gruzia. *B. cardui* Linné — Gruzia. *B. persicaecola* Boisduval — Gruzia. *B. prunicola* Kaltenbach — Corse (Starý et al. 1973). *B.* sp. — Gruzia. *Macchiatella* sp. — Italy (Starý 1966). *Myzus cerasi* Fabricius — Gruzia. *M. persicae* Sulzer — Italy (Starý 1966, 1967). *Toxoptera aurantii* Boyer de Fonscolombe — Italy (Starý 1964, 1966, 1967, Roberti 1969, Tremblay 1967). Gruzia (Starý 1968). *Without host data* — s. France (Starý et al. 1971). Spain (Quilis 1931, Chalver 1973). Gruzia.

Distr.: Algeria, Spain, s. France, Monaco, Corse, Italy, Sicily, Greece, Gruzia.

Biol.: Relation to ants in Italy (Starý 1966). Integrated control on Citrus in Italy (Starý 1964, 1966), Black Sea coastal zone of U.S.S.R, (Starý 1968).

Genus: *Lysaphidus* Smith 1944

L. arvensis Starý 1960

Hosts: *Coloradoa bournieri* Remaudière et Leclant — s. France (Starý et al. 1973). Distr.: s. France.

L. viaticus Sedlag 1959

Hosts: *Pleotrichophorus duponti* Hille Ris Lambers — s. France (Starý et al. 1971). Distr.: s. France.

L. sp.

Hosts: *Coloradoa moralesi* Remaudière et Leclant — Spain.

Genus: *Lysephedrus* Starý 1958

[**L. validus** (Haliday 1833)]

Hosts: *Aphis rumicis* Linné — Cyprus (Wilkinson 1926) (!). Distr.: Cyprus (?).

Genus: *Lysiphlebus* Förster 1862

Syn.: *Aphidaria* Provancher 1888. *Lysiphlebus* Förster subg. *Platycyphus* Mackauer 1960.

Subgenera: *Adialytus* Förster 1862. *Lysiphlebus* s. str., *Phlebus* Starý 1975.

21

L. (P.) ambiguus (Haliday 1834)

Syn.: ?*Aphidius* (*Aphidius*) *exiguus* Haliday 1834. ?*Aphidius diminuens* Nees 1834.

Hosts: *Aphis arbuti Ferrari* — Corse (Starý et al. 1971). *A. armata* Hausmann — s. France (Starý et al. 1973). *A. clematidis* Koch — s. France. *A. craccivora* Koch — Italy (Tremblay 1967). Iraq (Al — Azawí 1970, Starý 1969, Starý and Kaddou 1971). Gruzia. s. France. *A. fabae* Scopoli — Algeria. Corse (Starý et al. in press). Italy (Starý 1966, 1974, Roberti 1969, Tremblay 1967). Sicily (Starý 1966). Iraq (Starý and Kaddou 1971, Starý 1969, Al — Azawí 1970). Crimea (Starý 1965). Gruzia (Achvlediani 1964). *A. farinosa* Gmelin — Italy (Starý 1966, 1974, Mackauer 1960). Gruzia (Achvlediani 1964, Starý 1965). *A. gentianae* Börner — Crimea (Starý 1965). *A. gossypii* Glover — Corse (Starý et al. in press). Iraq (Starý 1969, Starý and Kaddou 1971, Al — Azawí 1970). Gruzia. *A. medicaginis* Koch — Gruzia. *A. nasturtii* Kaltenbach — Algeria. *A. nerii* Boyer de Fonscolombe — Italy (Starý 1966, 1974). Sicily (Starý 1966, 1974). Monaco (Starý 1974). Iraq (Al — Azawí 1970). *A. picridophila* Hille Ris Lambers — Corse (Starý et al. 1971). *A. punicae* Passerini — Italy (Starý 1966, 1974). *A. ruborum* Börner — Italy (Roberti 1969). Sicily (Starý 1966). Corse (Starý et al. 1973). Israel (Mackauer 1960). *A. sanguisorbae* Schrank — Corse (Starý et al. in press). *A. sarothamni* Franssen — Sicily (Starý 1966, 1974). *A. solanella* Theobald — Sicily (Barbagallo 1974, Starý 1966, 1974). *A. urticata* Fabricius — Gruzia (Starý 1965). *A. vallei* Stroyan et Hille Ris Lambers — Turkey. *A.* sp. — Algeria. s. France. Corse (Starý et al. in press). Italy (Starý 1966, 1974, Roberti 1969). Sicily (Starý 1966). Greece (Liste d'Ident. 1971). Iraq (Al — Azawí 1970, Starý and Kaddou 1971, Starý 1969). Gruzia (Achvlediani 1964, Starý 1963). *Brachycaudus cardui* Linné — Italy (Starý 1966, 1974). Sicily (Starý 1966). Corse (Starý et al. in press). Gruzia (Starý 1965, Achvlediani 1964). *B. mimeuri* Remaudière — s. France (Starý et al. 1971). *B. tragopogonis* Kaltenbach — Gruzia. *B.* sp. — Gruzia (Starý 1965). *Capitophorus hippophaes* Walker — Gruzia. *Chromaphis juglandicola* Kaltenbach — Sicily (Starý 1966, 1974). *Cryptosiphum cinae* Nevsky — Gruzia. *Macrosiphoniella pulvera* Walker — s. France (Starý et al. 1971). *M.* sp. — Gruzia. *Melanaphis donacis* Passerini — s. France (Sharma 1965). *Toxoptera aurantii* Boyer de Fonscolombe — Sicily (Starý 1966, 1974). Crete. Greece (Argyriou 1970). Israel (Rosen 1964, 1966, 1967, 1969, Avidov and Harpaz 1969, Starý 1967). *Without host data* — Canary Isl. (Mackauer 1962). Algeria. Italy. Egypt (Mackauer and Starý 1967). Turkey (Liste d'Ident. 1971). Crimea. Azerbaidjan (Starý 1965). Gruzia (Starý 1965).

Distr.: Canary Isl., Algeria, Spain, s. France, Monaco, Corse, Italy, Sicily, Crete, Greece, Turkey, Egypt, Israel, Iraq, Crimea, Azerbaidjan, Gruzia.

L. (A.) arvicola Starý 1961

Syn.: *Lysiphlebus mackaueri* Starý 1961. *Lysiphlebus* (*Lysiphlebus*) *crocinus* Mackauer 1962.

Hosts: *Sipha maydis* Passerini — s. France (Starý et al. 1971). Italy (Starý 1966,

Tremblay 1967). Sicily (Starý 1966). Bulgaria (Starý 1962). *S.* sp. — Spain (Mackauer 1962).

Distr.: Spain, s. France, Italy, Sicily, Bulgaria.

Biol.: Relation to ants in Italy (Starý 1966).

[*L. (L.) dissolutus* (Nees 1811) sensu Förster 1862]

Syn.: *Lysiphlebus (Platycyphus) macrocornis* Mackauer 1960.

Hosts: *Aphis rumicis* Linné — Italy (Menozzi 1937) (!).

Distr.: Italy (?).

L. (P.) fabarum (Marshall 1896)

Syn.: *Aphidius cardui* Marshall 1896. *Aphidius aurantii* Pierantoni 1907. *Aphidius gomezi* Quilis 1930. *Lysiphlebus fabarum* Marshall var. *inermis* Quilis 1931. *Lysiphlebus innovatus* Quilis 1931. *Aphidius janinii* Quilis 1930. *Lysiphlebus moroderi* Quilis 1931.

Hosts: *Aphis acetosae* Linné — s. France (Starý et al. 1971). *A. affinis* Del-Guercio — Iraq (Starý and Kaddou 1971). *A. arbuti* Ferrari — s. France (Starý et al. 1971). Corse (Starý et al. in press). *A. chloris* Koch — Italy (Starý 1966). Gruzia. *A. clematidis* Koch ·- Italy (Starý 1966). *A. confusa* Walker — s. France (Starý et al. 1973). *A. craccivora* Koch — Algeria. s. France (Starý et al. 1971). Corse (Starý et al. 1971). Italy (Starý 1966, 1967, Roberti 1969, Tremblay 1967). Sicily (Barbagallo 1974). Iraq (Starý and Kaddou 1971, Starý 1969). Gruzia. *A. cytisorum* Hartig — s. France (Starý et al. 1971). *A. dorycnii* Hille Ris Lambers — Corse (Starý et al. 1971). *A. fabae* Scopoli — Algeria. Portugal. s. France (Starý et al. 1971). Corse (Starý et al. 1971). Italy (Starý 1966, Roberti 1969). Sicily (Starý 1966, Mackauer 1960). Morocco. Bulgaria (Starý 1962). Iraq (Starý and Kaddou 1971, Starý 1969, Al — Azawí 1970). Crimea (Starý 1965). Gruzia (Achvlediani 1964, Starý 1965). *A. farinosa* Gmelin — Italy (Goidanich 1934, cf. Mackauer 1960) (?). *A. frangulae* Koch — Italy (Mackauer 1960). s. France. *A. fumanae* Remaudiere and Leclant — s. France (Starý et al. 1971). *A. gossypii* Glover — Algeria. Morocco (Mackauer 1960). Corse (Starý et al. 1971). Italy (Roberti 1969, Tremblay 1967). Bulgaria (Starý 1962, 1967, Radev 1968). Israel (Avidov and Harpaz 1969). Iraq (Al — Azawí 1970). *A. hederae* Kaltenbach — s. France (Starý et al. 1971). Gruzia. *A. idaei* van der Goot — Gruzia (Achvlediani 1964). *A. intybi* Koch — Corse (Starý et al. in press). Italy. *A. lambersi* Börner — Corse (Starý et al. in press). *A. medicaginis* Koch — Gruzia. *A. nasturtii* Kaltenbach — Algeria. *A. nerii* Boyer de Fonscolombe — s. France. *A. parietariae* Theobald — Italy (Tremblay 1967). *A. pomi* DeGeer — Corse (Starý et al. in press). *A. poterii* Börner — s. France. *A. punicae* Passerini — Algeria. *A. ruborum* Börner — Corse (Starý et al. in press). Sicily (Starý 1966). Italy (Roberti 1969). Iraq (Starý and Kaddou 1971). Gruzia. *A. rufula* Börner — Corse (Starý et al. 1973). *A. rumicis* Linné — Corse (Starý et al. in press). *A. salviae* Walker — Gruzia. *A. sanguisorbae* Schrank — s. France (Starý et al. 1973). *A. sarothamni* Franssen — s. France (Starý et al. 1973). Italy (Starý 1966). Sicily (Starý 1966). *A. solanella* Theobald — Portugal. Spain (Quilis 1931, Chalver 1973). Sicily (Barbagallo 1974). Greece (Zouliamis 1968). Iraq **23**

(Starý and Kaddou 1971). Gruzia. *A. spiraecola* Patch — Algeria. Corse (Starý et al. 1971). s. France. *A. tirucallis* Hille Ris Lambers — s. France (Starý et al. 1971). *A. umbrella* Börner — Italy (Starý 1966, Roberti 1969). *A. urticata* Fabricius — Italy (Starý 1966, Mackauer 1960). Gruzia. *A. verbasci* Schrank — Corse (Starý et al. 1973). *A. vitalbae* Ferrari — Corse (Starý et al. 1971). s. France (Starý et al. 1973). *A. viticis* Ferrari — Italy (Tremblay 1967). *A. sp.* — Algeria. s. France. Corse (Starý et al. 1971). Italy (Roberti 1969, Starý 1966). Sicily (Starý 1966). Monaco. Bulgaria (Starý 1962). Syria. Iraq (Starý and Kaddou 1971, Starý 1969). Crimea (Starý 1965). Gruzia (Achvlediani 1964, Starý 1965). *Brachycaudus cardui* Linné — s. France. Corse (Starý et al. in press). Sicily (Barbagallo 1974). Bulgaria (Starý 1962). Gruzia (Achvlediani 1964, Starý 1965). *B. helichrysi* Kaltenbach — Algeria. Italy (Tremblay 1967). *B. prunicola* Kaltenbach — Gruzia. *B. prunicola schwartzi* Börner — s. France (Starý et al. 1971). *B. rumexicolens* Patch — Gruzia. *B. tragopogonis* Kaltenbach — Gruzia. *B. sp.* — Iraq (Starý and Kaddou 1971). Yugoslavia. *Brachyunguis tamaricis* Lichtenstein — s. France (Starý et al. 1971). *B. tamaricophila* Nevsky — Gruzia. *Capitophorus carduinus* Walker — Iraq (Starý and Kaddou 1971). *C. eleagni* DelGuercio — Sicily (Barbagallo 1974). *C. inulae* Passerini — Italy (Tremblay 1967). *C. sp.* — Iraq (Starý 1969). *Cavariella aegopodii* Scopoli — Corse (Starý et al. in press). Spain (Quilis 1931). *Dysaphis cynarae* Theobald — Sicily (Barbagallo 1974). *D. plantaginea* Passerini — Italy. *D. sp.* — Turkey. *Hyalopterus pruni* Geoffroy — Algeria. *Macrosiphoniella sanborni* Gillette — Gruzia (Achvlediani 1964). *Melanaphis donacis* Passerini — s. France (Sharma 1965). *Myzus cerasi* Fabricius — Gruzia (Achvlediani 1964). *M. cerasi veronicae* Walker — s. France (Starý et al. 1971). *M. persicae* Sulzer — Gruzia. *Neanuraphis rhamni* Boyer de Fonscolombe — Gruzia. *Pemphigus lichtensteini* Tullgren — Gruzia (Achvlediani 1964, Starý 1965). *Protaphis* ? *terricola* Rondani — s. France (Starý et al. 1971). *P. sp.* — Corse (Starý et al. 1971). Italy (Starý 1966 — !, cf. *L. hispanus*). Turkey. *Rhopalosiphum maidis* Fitch — Morocco (Mackauer 1960). *Sitobion avenae* Fabricius — Spain (Quilis 1931, Chalver 1973) (?). *Toxoptera aurantii* Boyer de Fonscolombe — Italy (Pierantoni 1907). Turkey (Liste d'Ident. 1971). Israel (Rosen 1964, 1966, 1967, 1969, Starý 1967, Avidov and Harpaz 1969). Gruzia (Gaprindašvili 1956). *T. sp.* — Italy (Mackauer 1960). *Without host data* — Algeria (Mackauer and Starý 1967). Morocco (Mimeur 1934). Spain (Quilis 1930, 1931, Gómez Menor 1965). s. France (Starý et al. 1971, Biliotti and Sharma 1965). Italy. Turkey (Liste d'Ident. 1971). Egypt (Mackauer and Starý 1967). Lebanon (Mackauer and Starý 1967). Iraq (Liste d'Ident. 1971). Crimea (Starý 1965). Gruzia (Starý 1965). Azerbaidjan (Starý 1965).

Distr.: Spain, Portugal, Algeria, Morocco, s. France, Monaco, Corse, Italy, Sicily, Bulgaria, Greece, Turkey, Egypt, Syria, Lebanon, Israel, Iraq, Crimea, Gruzia, Azerbaidjan.

Biol.: Morphology (Goidanich 1934 ?, Tremblay 1964). Host range, biology, phenology in Italy (Tremblay 1964, Pierantoni 1907). Thelyotokous strain in Israel (Rosen 1967). Effectiveness in Italy (Tremblay 1964, Roberti 1969), Iraq (Al — Azawí 1970), Israel (Rosen 1967). Relation to ants in Italy (Starý 1966), Iraq (Starý 1969). Biological control on vegetables in France (Biliotti and Sharma 1965).

L. (P.) hispanus Starý 1973

Hosts: *Protaphis (Absintaphis)* sp. — Spain (Starý and Remaudière 1973). Italy (Starý 1966, as *L. fabarum*).
Distr.: Spain, Italy.

L. (A.) salicaphis (Fitch 1855)

Syn.: *Trioxys salicaphis* Fitch 1855. *Trioxys populaphis* Fitch 1855. *Lipolexis salicaphidis* Ashmead 1889. *Aphidius (Diaeretus) laticephalus* Telenga 1953.
Hosts: *Chaitophorus leucomelas* Koch — s. France (Starý et al. 1973). *C. niger* Mordwilko — s. France. Gruzia. *C. tremulae* Koch — Gruzia. *C.* sp. — s. France. Italy (Starý 1966, Liste d'Ident. 1971). Bulgaria (Starý 1962, Starý and Schlinger 1967). Iraq (Starý and Kaddou 1971). Gruzia.
Distr.: s. France, Italy, Bulgaria, Iraq, Gruzia.
Biol.: Relation to ants in Italy (Starý 1966).

L. (P.) testaceipes (Cresson 1880)

Syn.: ?*Praon viburnaphis* Fitch 1855. ?*Trioxys fuscatus* Cresson 1865. *Aphidius citraphis* Ashmead 1880. *Adialytus maidaphidis* Carman 1885. *Aphidius flavicoxa* Ashmead 1888. *Aphidaria basilaris* Provancher 1888. *Lysiphlebus minutus* Ashmead 1889. ?*Lysiphlebus piceiventris* Ashmead 1889. *Lysiphlebus cucurbitaphidis* Ashmead 1889. *Lysiphlebus eragrostaphidis* Ashmead 1889. *Lysiphlebus coquilleti* Ashmead 1889. *Lysiphlebus myzi* Ashmead 1889. *Lysiphlebus gossypii* Ashmead 1889. *Lysiphlebus abutilaphidis* Ashmead 1889. *Lysiphlebus tritici* Ashmead 1889. *Lysiphlebus persicaphidis* Ashmead 1889. *Lysiphlebus baccharaphidis* Ashmead 1889. *Aphidius persiaphis* Cook 1891. *Aphidius (Lysiphlebus) chrysoaphidis* Smith 1944.
Hosts: *Aphis craccivora* Koch — s. France (Starý et al. in press). *A. fabae* Scopoli — s. France (Starý et al. in press). *A. intybi* Koch — s. France (Starý et al. in press). *A. nerii* Boyer de Fonscolombe — s. France (Starý et al. in press). *A. rumicis* Linné — s. France (Starý et al. in press). *A. spiraecola* Patch — s. France (Starý et al. in press). Corse (Starý et al. in press). *Toxoptera aurantii* Boyer de Fonscolombe — s. France (Starý et al. in press). Corse (Starý et al. in press).
Distr.: s. France, Corse.
Biol.: Biological control on Citrus in s. France and Corse (Starý et al. in press).

L. (A.) thelaxis Starý 1961

Hosts: *Thelaxes dryophila* Schrank — Sicily (Starý 1966). Corse (Starý et al. in press). *T. suberi* DelGuercio — Corse (Starý et al. in press). Iraq (Starý and Kaddou 1971, Starý 1969). *T.* sp. — Turkey. *Without host data* — Italy (Tremblay 1967).
Distr.: Corse, Italy, Sicily, Turkey, Iraq.
Biol.: Notes — Iraq (Starý 1969). Relation to ants in Italy (Starý 1966), Iraq (Starý 1969).

L. sp.

Hosts: *Aphis fabae* Scopoli − Lebanon (Biol. Contr. Inf. Bull. 1967). *A. gossypii* Glover − Greece (Mentzelos et al. 1964). *Rhopalosiphum padi* Linné − Israel (Bodenheimer and Swirski 1957).

Genus: *Monoctonia* Starý 1962

M. pistaciaecola Starý 1962

Hosts: *Forda follicularia* Passerini − Gruzia (Starý 1968). *F. hirsuta* Mordwilko − Crimea (Starý 1962, 1965, 1968). *F.* sp. − Iraq (Starý and Kaddou 1971). *Pemphigus* sp. − Sicily (Starý 1966, 1968).
Distr.: Sicily, Iraq, Crimea, Gruzia.
Biol.: Seasonal occurrence, diapause (Starý 1966, 1970, Starý and Kaddou 1971).

Genus: *Monoctonus* Haliday 1833

Syn.: *Aphidileo* Rondani 1848.
Subgenera: *Monoctonus* s. str.

[**Mon. (M.) caricis** (Haliday 1833)]

Syn.: ?*Aphidius* (*Aphidius*) *fumatus* Haliday 1834.
Hosts: *Rhopalosiphum insertum* Walker − Italy (Vidano 1959; ! − to *M. cerasi*).
Distr.: Italy (?).

M. (M.) cerasi (Marshall 1896)

Syn.: ?*Aphidius cardui* Marshall var. *polygoni* Ivanov 1927. *Aphidius ivanovi* Mackauer 1967.
Hosts: *Rhopalosiphum insertum* Walker − Italy (Starý 1966). *Without host data* − Italy (Mackauer and Starý 1967).
Distr.: Italy.

M. (M.) crepidis (Haliday 1834)

Syn.: *Aphidius tuberculatus* Wesmael 1835. *Monoctonus paludum* Marshall 1896.
Hosts: *Nasonovia nigra* Hille Ris Lambers − s. France (Starý et al. 1973). Corse (Starý et al. 1971). *N. ribis − nigri* Mosley − s. France (Starý et al. 1971). *Without host data* − Spain (Chalver 1973).
Distr.: Spain, s. France, Corse.

Genus: *Paralipsis* Förster 1862

26 Syn.: *Myrmecobosca* Maneval 1940

P. enervis (Nees 1834)

Syn.: *Myrmecobosca mandibularis* Maneval 1940. *Myrmecobosca linnei* Hincks 1949.
Hosts: *Without host data* — Spain (Chalver 1973). Gruzia.
Distr.: Spain, Gruzia.

Genus: *Pauesia* Quilis 1931

P. abietis (Marshall 1896)

Hosts: *Cinara pinicola* Kaltenbach — Turkey (Schimitschek 1944). *Without host data* — Spain (Chalver 1973).
Distr.: Spain, Turkey.

P. antennata (Mukerji 1950)

Syn.: *Aphidius chloratus* Telenga 1953.
Hosts: *Pterochloroides persicae* Cholodkovsky — Iraq (Note: Starý and Kaddou 1971, mummies only, supposed to belong to *P. antennata*).
Distr.: Iraq.

P. cupressobii (Starý 1960)

Hosts: *Cinara juniperi* DeGeer — Corse (Starý et al. 1973).
Distr.: Corse.

P. goidanichi Starý 1966

Hosts: *Cinara juniperi* DeGeer — Italy. *C.* sp. — Italy (Starý 1966).
Distr.: Italy.

P. infulata (Haliday 1834)

Syn.: *Paraphidius albiflagellaris* Starý 1960.
Hosts: *Cinara pectinatae* Nordlinger — s. France (Starý et al. 1971).
Distr.: s. France.

P. juniperorum (Starý 1960)

Hosts: *Cinara juniperi* DeGeer — Corse (Starý et al. in press). *C.* sp. — Italy (Starý 1966).
Distr.: Italy, Corse.

P. laricis (Haliday 1834)

Hosts: *Cinara* sp. — Italy (Starý 1966).
Distr.: Italy.

27

P. piceaecollis (Starý 1960)

Hosts: *Cinara maghrebica* Mimeur — Sicily (Tremblay 1970). *C.* sp. — Italy (Tremblay 1967).
Distr.: Italy, Sicily.

P. picta (Haliday 1834)

Hosts: *Cinara pini* Linné — Bulgaria. *C. schimitscheki* Börner — Italy (Tremblay 1970).
Distr.: Italy, Bulgaria.

P. pini (Haliday 1934)

Syn.: *Aphidius planistipes* Nees 1834. ?*Aphidius varius* Nees 1834. ?*Aphidius panzerii* Rondani 1848. *Aphidius lachnivorus* Ashmead 1906.
Hosts: *Cinara excelsae* Hille Ris Lambers — Corse (Starý et al. in press). *C. palaestinensis* Hille Ris Lambers — Israel (Bodenheimer and Swirski 1957, Bodenheimer and Neumark 1955, Avidov and Harpaz 1969). *C. schimitscheki* Börner -- Sicily (Tremblay 1970). *C.* sp. — Italy (Starý 1966). *Without host data* — Spain (Chalver 1973).
Distr.: Spain, Corse, Italy, Sicily, Israel.
Biol.: Israel (Bodenheimer and Neumark 1955).

P. pinicollis (Starý 1960)

Hosts: *Without host data* — Spain (Chalver 1973).
Distr.: Spain.

P. rufiabdominalis Starý 1966

Hosts: *Cinara excelsae* Hille Ris Lambers — Italy. Corse (Starý et al. in press). *C.* sp. — Italy (Starý 1966).
Distr.: Corse, Italy.

P. silana Tremblay 1969

Hosts: *Cinara acutirostris* Hille Ris Lambers — Italy (Tremblay 1969). Sicily (Tremblay 1970). *C. maghrebica* Mimeur — s. France (Starý et al. 1973). *C. pini* Linné — Italy.
Distr.: s. France, Italy, Sicily.

P. silvestris (Starý 1960)

Hosts: *Cinara excelsae* Hille Ris Lambers — Italy (Starý 1966). Sicily (Starý 1966). *C. pini* Linné — Italy (Starý 1966). *C.* sp. — Italy (Starý 1966).
Distr.: Italy, Sicily.
28   Biol.: Relation to ants in Italy (Starý 1966).

P. unilachni (Gahan 1926)

Syn.: *Pauesia albuferensis* Quilis 1931. *Aphidius praevisus* Gautier et Bonnamour 1936. *Trioxys basilewskyi* Benoit 1955.

Hosts: *Schizolachnus pineti* Fabricius — Italy (Tremblay 1970, Schimitschek 1967). Sicily (Tremblay 1970). Bulgaria. *S.* sp. — Italy (Starý 1966, Tremblay 1967). *Cinara nuda* Mordwilko — Spain (Quilis 1931) (!). *Without host data* — Spain (Chalver 1973).

Distr.: Spain, Italy, Sicily, Bulgaria.

P. sp.

Hosts: *Cinara juniperi* DeGeer — Corse (Starý et al. in press).

Genus: *Praon* Haliday 1833

Syn.: *Achoristus* Ratzeburg 1852. *Aphidaria* Provancher 1886.

P. abjectum (Haliday 1833)

Syn.: *Bracon (Achoristus) aphidiiformis* Ratzeburg 1852. ?*Praon peregrinus* Ruthe 1859.

Hosts: *Aphis craccivora* Koch — Iraq (Starý 1969, Starý and Kaddou 1971). *A. fabae* Scopoli — Sicily (Starý 1966). *A. punicae* Passerini — Iraq (Starý and Kaddou 1971). *A. ruborum* Börner — Corse (Starý et al. in press). *A. sambuci* Linné — s. France (Starý et al. 1973). *A. solanella* Theobald — Iraq (Starý and Kaddou 1971). *A. spiraecola* Patch — Corse (Starý et al. 1973). *A. viticis* Ferrari — Iraq (Starý and Kaddou 1971). *A.* sp. — Bulgaria. Iraq (Starý 1969, Starý and Kaddou 1971). *Rhopalosiphum padi* Linné — Italy. *Without host data* — Spain (Chalver 1973). Near East (Mackauer 1962).

Distr.: Spain, s. France, Corse, Italy, Sicily, Bulgaria, Iraq.

Biol.: Relation to ants in Italy (Starý 1966), Iraq (Starý 1969).

P. absinthii Bignell 1894

Hosts: *Macrosiphoniella absinthii* Linné — Italy. *M. artemisiae* Boyer de Fonscolombe — Gruzia. *M. leucanthemi* Ferrari — s. France (Starý et al. 1973). *M. sanborni* Gillette — Gruzia. *M. staegeri* Hille Ris Lambers — s. France (Starý et al. 1971).

Distr.: s. France, Italy, Gruzia.

P. barbatum Mackauer 1967

Hosts: *Acyrthosiphon pisum* Harris — Italy (Starý 1966, as *P. dorsale*). Gruzia. *Without host data* — Morocco, Spain.

Distr.: Spain, Morocco, Italy.

29

P. bicolor Mackauer 1959

Hosts: *Eulachnus rileyi* Will. − Sicily (Tremblay 1970). *Without host data* − Spain (Mackauer and Starý 1967).
Distr.: Spain, Sicily.

P. dorsale (Haliday 1833)

Syn.: *Blacus discolor* Nees 1834. *Praon longicorne* Marshall 1896.
Hosts: *Acyrthosiphon pisum* Harris − Italy (Starý 1966, to *P. barbatum*). *Staticobium limonii* Mordwilko − Corse (Starý et al. in press). *Uroleucon carthami* Hille Ris Lambers − Sicily (Starý 1966). Crimea (Starý 1965). *U. chondrillae* Nevsky − Corse (Starý et al. 1973). *U. cichorii* Koch − s. France (Starý et al. 1971). Corse (Starý et al. in press). Bulgaria (Starý 1962). *U. jaceae* Linné − Corse (Starý et al. 1971). Sicily (Starý 1966). *U. sonchi* Geoffroy − Bulgaria (Starý 1962). *U. sp.* − Corse (Starý et al. in press). Italy (Roberti 1969). *Without host data* − Spain (Chalver 1973). Italy (Mackauer and Starý 1967).
Distr.: Spain, s. France, Corse, Italy, Sicily, Bulgaria.

P. exsoletum (Nees 1811)

Syn.: *Praon palitans* Muesebeck 1956.
Hosts: *Therioaphis ononidis* Kaltenbach − s. France (Starý et al. 1973). Italy (v. d. Bosch 1957). *T. trifolii* Monell − Italy (Starý 1966, v. d. Bosch 1957). Bulgaria (Pělov 1972). Turkey (v. d. Bosch 1957). Lebanon (v. d. Bosch 1957). Israel (Avidov and Harpaz 1969, Harpaz 1955 − as *Praon* sp., v. d. Bosch 1957, Mackauer 1959). *T.* sp. − Iraq (Starý and Kaddou 1971). Israel (Muesebeck 1956). *Without host data* − Egypt (Mackauer 1959), Israel (Mackauer 1959), Spain (Chalver 1973).
Distr.: Spain, s. France, Italy, Bulgaria, Turkey, Lebanon, Israel, Iraq.
Biol.: Effectiveness in Israel (Harpaz 1955).

P. flavinode (Haliday 1833)

Syn.: *Blacus emacerator* Nees 1834. *Praon glabrum* Starý et Schlinger 1967.
Hosts: *Tuberculoides albosiphonatus* Hille Ris Lambers − Iraq. *T. moerickei* Hille Ris Lambers − Iraq. *T.* sp. − Iraq (Starý and Kaddou 1971). *Without host data* − Spain (Chalver 1973).
Distr.: Spain, Iraq.

P. grossum Starý 1971

Hosts: *Amphorophora rubi* Kaltenbach − s. France (Starý et al. 1973). *Aulacorthum solani* Kaltenbach − s. France (Starý et al. 1973).

30   Distr.: s. France.

P. myzophagum Mackauer 1959

Hosts: *Myzus persicae* Sulzer — s. France (Starý et al. 1971). Mediterranean (Mackauer 1962). *Without host data* — Spain (Mackauer and Starý 1967, Chalver 1973). Israel (Mackauer and Starý 1967).
Distr.: Spain, s. France, Israel.

P. necans Mackauer 1959

Hosts: *Without host data* — Iraq (Liste d'Ident. 1971).
Distr.: Iraq.

P. rosaecola Starý 1961

Hosts: *Macrosiphum rosae* Linné — Corse (Starý et al. in press). *Sitobion* sp·
— Corse (Starý et al. in press).
Distr.: Corse.

P. silvestre Starý 1971

Hosts: *Periphyllus* sp. — s. France, Bulgaria.
Distr.: s. France, Bulgaria.

P. volucre (Haliday 1833)

Syn.: *Blacus angulator* Nees 1834. *Aphidius aphidivorus* Ratzeburg 1844. *Praon pruni* Ivanov 1925.
Hosts: *Acyrthosiphon lambersi* Leclant et Remaudière — Corse (Starý et al. 1973). *A. pelargonii geranii* Kaltenbach — Sicily (Starý 1966, 1974). *A. pisum* Harris — Gruzia. *A.* sp. — Corse (Starý et al. in press). Sicily. *Aphis arbuti* Ferrari — s. France (Starý et al. 1971). *A. craccivora* Koch — Corse (Starý et al. 1971). *A. fabae* Scopoli — Greece (Zouliamis 1968). *A. pomi* DeGeer — Italy (Starý 1966, 1974). *A. solanella* Theobald — Greece (Zouliamis 1968). *A.* sp. — Italy (Starý 1966, 1974). Greece (Zouliamis 1968). *Aulacorthum solani* Kaltenbach — Sicily (Starý 1966). *Brachycaudus amygdalinus* Schouteden — Gruzia. *B. lychnidis* Linné — Corse (Starý et al. in press). s. France (Starý et al. 1973). *B.* sp. — Gruzia. *Eucarazzia elegans* Ferrari — s. France (Starý et al. 1971). *Hyalopterus pruni* Geoffroy — Corse (Starý et al. 1971). Italy (Starý 1966, 1964, Tremblay 1967). Yugoslavia (Starý 1974). Bulgaria (Starý 1974, Pělov 1972). Iraq (Starý and Kaddou 1971). Crimea (Starý 1962). Gruzia (Achvlediani 1964). Azerbaidjan (Starý 1965). *Hyalopterus* sp. — Italy (Quilis 1932, ?). *Hyperomyzus lactucae* Linné — Sicily (Starý 1966, 1974), Corse (Starý et al. in press). Greece (Zouliamis 1968). Azerbaidjan (Starý 1965). *H. picridis* Börner — s. France. *H. sp.* — Italy (Starý 1966, Tremblay 1967). Sicily (Starý 1966). *Macrosiphum euphorbiae* Thomas — Sicily (Starý 1966). Italy (Starý 1967, 1974, Roberti 1969). *M. rosae* Linné — Portugal. Italy (Starý 1966, 1974, Roberti 1969). Sicily. Yugoslavia. Greece (Zouliamis 1968). Iraq (Starý and Kaddou 1971). Crimea **31**

(Starý 1965). Gruzia. *Melanaphis donacis* Passerini — Iraq (Starý and Kaddou 1971). *Myzus persicae* Sulzer — Algeria. Corse (Starý et al. 1973). Israel (Starý 1967, Rosen 1964, 1966, 1967, 1969, Avidov et Harpaz 1969). Gruzia. *Rhopalosiphum padi* Linné — Israel (Mackauer 1959). *Sitobion avenae* Fabricius — Corse (Starý et al. in press). France (Starý et al. 1973). Italy (Starý 1966, Roberti 1969). *S. fragariae* Walker — Corse (Starý et al. in press). Israel (Mackauer 1959). Crimea (Starý 1965). *S.* sp. — Corse (Starý et al. in press). Bulgaria (Starý 1974). *Uroleucon jaceae* Linné — Italy (Starý 1966, 1974). *U. ochropus* Hille Ris Lambers — Crimea (Starý 1965). *U. sonchi* Geoffroy — Greece (Zouliamis 1968). *U.* sp. — Sicily (Starý 1966, 1974). *Without host data* — Spain (Chalver 1973). Algeria. Canary Isl. (Mackauer 1962). Italy (Mackauer 1968). Israel (Fleschner 1963). Crimea (Starý 1962, 1965). Azerbaidjan (Aliev 1971).

Distr.: Algeria, Canary Isl., Spain, Portugal, s. France, Corse, Italy, Sicily, Yugoslavia, Bulgaria, Greece, Israel, Iraq, Crimea, Gruzia, Azerbaidjan.

Biol.: Relation to ants in Italy (Starý 1966). Integrated control on peach in Italy (Starý 1964, 1966).

P. sp.

Hosts: *Aphis fabae* Scopoli — Corse (Starý et al. in press). *A.* sp. — Bulgaria (Starý 1962). *Aulacorthum solani* Kaltenbach — Bulgaria (Starý 1962). *Brachycaudus* sp. — Corse (Starý et al. 1973). *Chaitophorus* sp. — Iraq (Starý and Kaddou 1971). *Corylobium avellanae* Schrank — s. France. *Dysaphis devecta* Walker — Gruzia (Achvlediani 1964). *Hyalopterus* sp. — Bulgaria (Starý 1962). *Linosiphon galiophagus* Wimsh. — Corse (Starý et al. in press). *Macrosiphum rosae* Linné — s. France (Starý et al. 1973). *Myzus persicae* Sulzer — Corse (Starý et al. in press). *Therioaphis trifolii* Monell — Israel (Harpaz 1955). *Toxoptera aurantii* Boyer de Fonscolombe — Cyprus (Wood 1963). *Uroleucon cichorii* Koch — Corse (Starý et al. in press). s. France (Starý et al. 1973). *U.* sp. — s. France. *Without host data* — Israel (Bodenheimer and Swirski 1957).

Genus: **Protaphidius** Ashmead 1900

Syn.: *Coelonotus* Förster 1862. *Menozzia* Goidanich 1934.

P. wissmannii (Ratzeburg 1848)

Syn.: *Coelonotus rufus* Förster 1862. *Menozzia formicaria* Goidanich 1934.
Hosts: *Stomaphis* sp. — s. France. Italy (Starý 1966). *Without host data* — Italy (Goidanich 1934, Grandi 1951, Rondani 1874).
Distr.: s. France, Italy.
Biol.: (Goidanich 1934). (Starý 1970). Relation to ants (Goidanich 1934, Starý 1966, 1970).

Genus: *Trioxys* Haliday 1833

Syn.: *Misaphidus* Rondani 1848. *Nevropenes* Provancher 1886.
Subgenera: *Betuloxys* Mackauer 1960. *Binodoxys* Mackauer 1960. *Trioxys* s. str.

T. (B.) acalephae (Marshall 1896)

Syn.: *Trioxys amoplanus* Quilis 1934. *Trioxys* (*Trioxys*) *urticae* Mackauer 1959. *Trioxys* (*Trioxys*) *rietscheli* Mackauer 1959.

Hosts: *Aphis affinis* DelGuercio — Iraq (Starý and Kaddou 1971). *A. cisticola* Leclant et Remaudière — Corse (Starý et al. in press). *A. craccivora* Koch — Corse (Starý et al. in press). Gruzia. *A. cytisorum* Hartig — Bulgaria. *A. fabae* Scopoli — Italy (Tremblay 1967). *A. farinosa* Gmelin — Gruzia. *A. grossulariae* Kaltenbach — Turkey. *A. paralios* Hille Ris Lambers — s. France (Starý et al. 1971). *A. parietariae* Theobald — s. France (Starý et al. 1971). *A. ruborum* Börner — Corse (Starý et al. 1973). Italy (Starý 1966). *A. urticata* Fabricius — Italy. Gruzia. *A. vallei* Stroyan and Hille Ris Lambers — Turkey. *A.* sp. — s. France. Bulgaria. Iraq. Gruzia. *Without host data* — Morocco.
Distr.: Morocco, s. France, Corse, Italy, Bulgaria, Turkey, Iraq, Gruzia.
Biol.: Relation to ants in Italy (Starý 1966).

T. (T.) acericola Starý and Mackauer 1971

Hosts: *Drepanosiphoniella aceris fugans* Remaudière et Leclant — s. France (Starý and Mackauer 1971, Starý et al. 1971).
Distr.: s. France.

T. (B.) angelicae (Haliday 1833)

Syn.: *Trioxys placidus* Gautier 1922. *Trioxys granatensis* Quilis 1931. *Trioxys boscai* Quilis 1931. *Trioxys fumariae* Quilis 1931. *Trioxys obscuriformis* Quilis 1931. *Trioxys* (*Binodoxys*) *angelicae mediterraneus* Mackauer 1960.

Hosts: *Aphis arbuti* Ferrari — s. France (Starý et al. 1971, 1973). Corse (Starý et al. 1971). *A. cisticola* Leclant et Remaudière — Corse (Starý et al. in press). *A. craccivora* Koch — s. France. Italy (Tremblay 1967). Israel (Avidov and Harpaz 1969). Iraq (Starý and Kaddou 1971). Gruzia. *A. fabae* Scopoli — Algeria. s. France. Corse (Starý et al. 1971). Italy (Starý 1966, Roberti 1969, Tremblay 1967). Sicily (Starý 1966). Greece (Zouliamis 1968). Turkey (Liste d'Ident. 1971). Lebanon (Biol. Contr. Inf. Bull. 1967). Iraq (Al — Azawi 1970). *A. frangulae* Kaltenbach — Corse (Starý et al. in press). *A. gossypii* Glover — Italy (Roberti 1969). Morocco (Mimeur 1934). Corse (Starý et al. in press). Israel (Rosen 1964, 1966, 1967, 1969, Avidov and Harpaz 1969, Mackauer 1960). Iraq (Al — Azawi 1970). *A. hederae* Kaltenbach — s. France. Corse (Starý et al. in press). Italy (Starý 1966). Monaco. *A. lichtensteini* Leclant et Remaudière — s. France (Starý et al. 1971). *A. nerii* Boyer de Fonscolombe — Algeria. s. France. *A. paralios* Hille Ris Lambers — s. France (Starý et al. 1971). **33**

A. parietariae Theobald — s. France (Starý et al. 1971). *A. pomi* DeGeer — Italy (Starý 1966, Principi et. al. 1967, 1969, Liste d'Ident. 1971). Bulgaria (Pělov 1972). Azerbaidjan. Gruzia. *A. punicae* Passerini — Egypt. Iraq (Starý and Kaddou 1971, Al — Azawí 1970, Starý 1969). *A. ruborum* Börner — Corse (Starý et al. in press). Italy (Starý 1966). *A. rumicis* Linné — Spain (Mackauer 1960). *A. sambuci* Linné — Italy (Starý 1966, Tremblay 1967). *A. solanella* Theobald — s. France. Italy (Starý 1966). Greece (Zouliamis 1968). Iraq (Starý and Kaddou 1971). *A. spiraecola* Patch — s. France (Starý et al. 1971, Liste d'Ident. 1971). Corse (Starý et al. 1971). *A. umbrella* Börner — s. France (Starý et al. 1971). Greece. Italy (Roberti 1969). Corse (Starý et al. in press). *A. viticis* Ferrari — Iraq (Starý and Kaddou 1971). *A. zizyphi* Theobald — Iraq (Starý and Kaddou 1971). *A.* sp. — Algeria. s. France. Italy (Starý 1966). Morocco (Mimeur 1934). Iraq (Starý 1969). *Brachycaudus persicae niger* Smith — Italy (Tremblay 1967). *B. persicae* Passerini — Corse (Starý et al. in press). *B.* sp. — Corse (Starý et al. in press). *Dysaphis cynarae* Theobald — Sicily (Barbagallo 1974). *D. plantaginea* Passerini — Italy (Principi et al. 1967, Liste d'Ident. 1967). *Melanaphis pyraria* Passerini — Corse (Starý et al. in press). *Myzus persicae* Sulzer — Corse (Starý et al. 1973). *Toxoptera aurantii* Boyer de Fonscolombe — s. France. Italy (Starý 1967, 1969, Roberti 1969, Tremblay 1967). Sicily (Starý 1964, 1966). Israel (Rosen 1964, 1966, 1967, 1969, Starý 1967, 1969, Avidov and Harpaz 1969). — *"Apion arrogans"* (Col. Curculionidae) — Israel (Liste d'Ident. 1971) (!). *Without host data* — Algeria. Canary Isl. (Mackauer 1962). Tunisia (Mackauer and Starý 1967). s. France (Starý et al. 1971). Spain (Quilis 1931, Chalver 1973). Egypt (Mackauer and Starý 1967). Lebanon (Mackauer and Starý 1967). Israel (Fleschner 1963). Iraq (Liste d'Ident. 1971).

Distr.: Spain, Canary Isl., Algeria, Morocco, Tunisia, s. France, Monaco, Corse. Italy, Sicily, Greece, Egypt, Lebanon, Israel, Iraq, Gruzia, Azerbaidjan.

Biol.: Seasonal history in Italy (Principi et al. 1967). Relation to ants in Italy (Starý 1966). Iraq (Starý 1969). Effectiveness in Iraq (Al — Azawí 1970). Italy (Principi et al. 1967, 1969). Integrated control on apple in Italy (Principi et al. 1967, 1969), on Citrus in Italy (Starý 1964, 1966), in Black Sea coastal area of U.S.S.R. (Starý 1968).

T. (T.) auctus (Haliday 1833)

Hosts: *Rhopalosiphum insertum* Walker — Italy (Vidano 1959, 1963). *R. padi* Linné — Italy.
Distr.: Italy.

T. (B.) brevicornis (Haliday 1833)

Syn.: *Aphidius (Trioxys) minutus* Haliday 1833.
Hosts: *Cavariella aegopodii* Scopoli — Algeria. *Hyadaphis foeniculi* Passerini — s. France. Corse (Starý et al. in press). Turkey. Israel (Mackauer 1959). *Staegeriella necopinata* Börner — Corse (Starý et at. 1971). *Without host data* — Canary Isl. (Mackauer 1962). N. Africa (Mackauer 1962).

34   Distr.: Algeria, Canary Isl., s. France, Corse, Turkey, Israel.

T. (B.) centaureae (Haliday 1833)

Hosts: *Aphis rumicis* Linné − Cyprus (Wilkinson 1926) (!) (cf. Mackauer and Starý 1967, to *T. angelicae*). *Uroleucon jaceae* Linné − Italy (Starý 1966). *U.* sp. − Algeria. Gruzia.
Distr.: Algeria, Italy, Cyprus (?), Gruzia.

[*T. (T.) cirsii* (Curtis 1831)]

Syn.: *Aphidius (Trioxys) aceris* Haliday 1833. ?*Aphidius resolutus* Nees 1834.
Hosts: *Periphyllus* sp. − Italy (Starý 1966) (!This record should correctly be placed to *T. falcatus*).
Distr.: Italy (?).
Biol.: Relation to ants in Italy (Starý 1966).

T. (T.) complanatus Quilis 1931

Syn.: *Trioxys (Trioxys) utilis* Muesebeck 1956.
Hosts: *Therioaphis langloisi* Remaudière and Leclant − s. France (Starý et al. 1971). *T. littoralis* Hille Ris Lambers − Corse (Starý et al. 1973). *T. riehmi* Börner − Corse (Starý et al. 1971). *T. trifolii* Monell − Morocco. Italy (v. d. Bosch 1957). Turkey (v. d. Bosch 1957). Israel (Avidov and Harpaz 1969, v. d. Bosch 1957, Mackauer 1959). Iraq (v. d. Bosch 1957). Middle East (Muesebeck 1956). *T.* sp. − Spain (Muesebeck 1956). Morocco. Iraq (Al − Azawí 1970). *Without host data* − Spain (Quilis 1931).
Distr.: Spain, Morocco, s. France, Corse, Italy, Turkey, Israel, Iraq.
Biol.: Effectiveness in Iraq (Al − Azawí 1970).

T. (T.) crudelis (Rondani 1848)

This is an unclear species of *Trioxys*.

T. (T.) curvicaudus Mackauer 1967

Hosts: *Eucallipterus tiliae* Linné − Spain (Starý and Remaudière 1973).
Distr.: Spain.

T. (T.) falcatus Mackauer 1959

Hosts: *Periphyllus* sp. − Italy (Starý 1966, as *T. cirsii*). Gruzia.
Distr.: Italy, Gruzia.

T. (B.) heraclei (Haliday 1833)

Syn.: *Aphidius (Trioxys) letifer* Haliday 1833. *Aphidius obsoletus* Wesmael 1835.
Hosts: *Cavariella theobaldi* Gillette et Bragg − Italy. *C.* sp. − Gruzia. *Without host data* − Spain (Mackauer and Starý 1967).
Distr.: Gruzia, Spain, Italy.

35

T. (B.) hortorum Starý 1961

Hosts: *Myzocallis carpini* Koch — Gruzia.
Distr.: Gruzia.

T. (T.) humuli Mackauer 1960

Hosts: *Phorodon cannabis* Passerini — Gruzia. *P. humuli* Schrank — Corse (Starý et al. 1973).
Distr.: Corse, Gruzia.

T. (T.) pallidus (Haliday 1833)

Syn.: *Aphidius callipteri* Marshall 1896. *Trioxys pulcher* Gautier et Bonnamour 1924.
Hosts: *Chromaphis juglandicola* Kaltenbach — France (Fisher et al. 1959, Schlinger 1960, v. d. Bosch et al. 1962). Iraq (Starý and Kaddou 1971, Starý 1969). *Eucallipterus tiliae* Linné — Gruzia. *Hoplocallis picta* Ferrari — s. France (Starý et al. 1973). *Myzocallis carpini* Koch — Gruzia. *M. coryli* Goetze — Corse (Starý et al. in press). Italy (Starý 1966). *Tuberculoides albosiphonatus* Hille Ris Lambers — Iraq. *T. annulatus* Hartig — Sicily (Starý 1966). *T. moerickei* Hille Ris Lambers — Iraq. *T.* sp. — Italy (Starý 1966). Sicily (Starý 1966). Turkey. Iraq (Starý and Kaddou 1971, Starý 1969). *Without host data* — Spain (Mackauer and Starý 1967, Chalver 1973). Italy (Mackauer and Starý 1967).
Distr.: Spain, s. France, Corse, Italy, Sicily, Turkey, Iraq, Gruzia.

T. (T.) pannonicus Starý 1960

Hosts: *Titanosiphon artemisiae* Koch — Italy (Tremblay 1972). *Without host data* — Canary Isl. (Mackauer 1962).
Distr.: Canary Isl., Italy.
Biol.: Notes on distribution (Mackauer 1962, Starý 1970, Tremblay 1972).

T. (T.) phyllaphidis Mackauer 1961

Hosts: *Phyllaphis fagi* Linné — Gruzia.
Distr.: Gruzia.

T. (T.) quercicola Starý 1969

Hosts: *Thelaxes suberis* Del Guercio — Iraq (Starý and Kaddou 1971, Starý 1969).
Distr.: Iraq.

T. sp.

Hosts: *Aphis craccivora* Koch — Gruzia (Achvlediani 1964). *A. parietariae* Theobald — Iraq (Starý and Kaddou 1971). *A.* sp. — Greece (Mentzelos et al. 1964). Iraq (Starý and Kaddou 1971). *Brachycaudus* sp. — Iraq (Starý and Kaddou 1971). *Capitophorus eleagni* DelGuercio — s. France (Starý et al. 1973). *Cavariella* sp. — s. France. *Toxoptera aurantii* Boyer de Fonscolombe — Cyprus (Wood 1963). *Without host data* — Morocco, Turkey. Iraq (Liste d'Ident. 1971). Gruzia.

II. DISTRIBUTION

Characteristics of the area

The Mediterranean subregion of the Palearctic region covers approximately the following area: the Canary Islands; Portugal; Spain; the mediterranean France; Monaco; the Baleares; Corse; Sardegna; Italy (with the exception of the Alps); Sicily; North Africa (north of the Tropic of Cancer, i.e. Morocco, Algeria, Tunisia, Libya, U.A.R. — Egypt); the mediterranean coastal areas of Yugoslavia and Albania; most of the lowland and Black Sea coastal areas of Bulgaria; Greece; Turkey; Cyprus; Asia Minor (Saudi Arabia — north of the Tropic of Cancer, Lebanon, Israel, Syria, Iraq, Kuwait); the Crimea; the Transcaucasia (including the Black Sea and Caspian coastal areas); Caspian coastal areas of Iran.

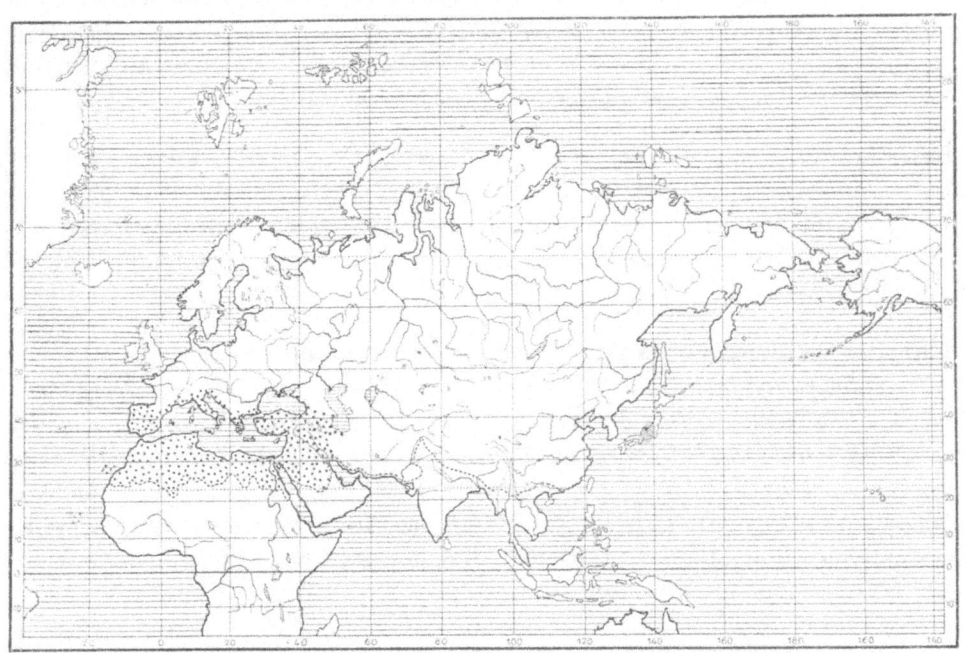

Map 1. The Mediterranean area.

The Mediterranean area cannot be strictly defined as there are widely transitional areas particularly with the neighbouring Eurosiberian and Central Asian subregions of the Palearctics.

Analysis of the parasite fauna

The analysis of the parasite fauna of the Mediterranean that has included almost 100 ascertained species indicates that the main features have become known. According to the classification of the world parasite fauna by STARÝ (1968, 1970) the following faunistic complexes (abbreviation "FC") can be distinguished:

Holarctic Forest Tundra FC. — *Aphidius cingulatus, Lysiphlebus salicaphis*. It is typical of the cool conditions of arctic climate, being a transitional zone between forest-free tundra area in the north and mostly coniferous taiga forests in the south. Many elements of the forest tundra may be found far in the south, in parks, forests, river valleys, peat-bogs, and at higher altitudes in the mountains.

Boreal Europe FC. — *Praon necans*. This complex is typical of cooler climatic conditions. Because of its origin, it is probably closely connected with forest-tundra zone. It is restricted to the northern parts of Europe, separate species being distributed to the south in a similar way as in the forest-tundra elements and may be found in peat-bogs, mountain and submountain forest undergrowth, etc.

West Eurasian Coniferous Forest FC. — *Pauesia cupressobii, P. goidanichi, P. juniperorum, P. piceaecollis, P. picta, P. pinicollis, P. silvestris, Praon bicolor*. It is typical of coniferous and mixed forest of Europe. Members of this complex are distributed almost all over Europe, penetrating to the Mediterranean.

East Eurasian Coniferous Forest FC. — *Diaeretus leucopterus, Pauesia abietis, P. infulata, P. laricis, P. pini, P. unilachni*. It is typical of the Far Eastern type of coniferous forest. Some species of this complex are also widely distributed in Europe. Their distribution areas are either disjunct or may be almost continuous but poorly known.

European Deciduous Forest FC. — *Aphidius hieraciorum, A. hortensis, Areopraon lepelleyi, Ephedrus cerasicola, E. minor, Lysiphlebus ambiguus, L. dissolutus, L. thelaxis, Monoctonus caricis, M. cerasi, M. crepidis, Praon abjectum, P. flavinode, P. grossum, P. rosaecola, P. silvestre, P. volucre, ?Protaphidius wissmannii, Trioxys angelicae, T. heraclei, T. cirsii, T. curvicaudus, T. falcatus, T. hortorum, T. humuli, T. pallidus, T. phyllaphidis*. It is typical of the deciduous forests of Europe. All the species are mostly restricted to Europe in distribution, some of them penetrate as far as the Mediterranean and Central Asia.

Far Eastern Deciduous Forest FC. — *Aphidius salicis, Ephedrus lacertosus, E. persicae, E. plagiator, Lysephedrus validus*. It is typical of the Far Eastern deciduous forest type; some species penetrate to the allied tropics, or, on the other hand, exhibit transpalearctic distribution, their areas being either disjunct or almost continuous.

Mediterranean Forest and Shrub FC. — This is a newly defined and distinguished complex. It is associated with evergreen sclerophylous and coniferous forest and shrub formations of the Mediterranean. The following species are included here: *Pauesia silana, P. rufiabdominalis* (formerly included in the West European Coniferous Forest FC), *Trioxys acericola, Trioxys quercicola, Monoctonia pistaciaecola* (formerly included in the Mediterranean FC). Some members penetrate to Central Asia and to the other parts of Europe: *Monoctonia pistaciaecola* is a parasite of *Forda* species that form galls on *Pistacia,* a tertiary relic in the Mediterranean and in Central Asia, and of *Pemphigus* sp. that form galls on *Populus.* As *Pistacia* have been eradicated in Europe north of the Mediterranean in the course of the Quaternary, the *Forda* species are distributed here in anholocyclic progeny and occur on the roots of grasses and are parasitized here by other parasite species. However, *M. pistaciaecola* is distributed also in this area (up to the British Isles) and parasitizes *Pemphigus* species (see STARÝ 1968, 1970).

Eurasian Steppes FC. — *Aphidius absinthii, A. ervi, A. funebris, A. matricariae, A. phalangomyzi, A. picipes, A. sonchi, A. uzbekistanicus, Diaeretiella rapae, Ephedrus nacheri, E. niger, Lipolexis gracilis, Lysaphidus arvensis, L. viaticus, Lysiphlebus arvicola, L. fabarum, Paralipsis enervis, Praon absinthii, P. barbatum, P. dorsale,*

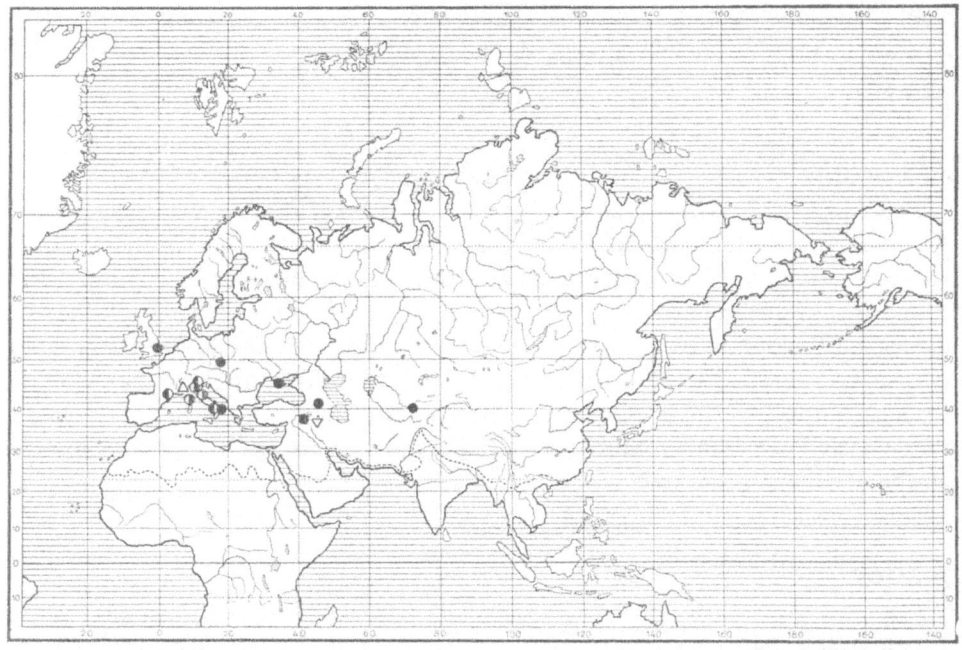

Map 2. Mediterranean forest and shrub FC. ◑ *Pauesia silana,* ◐ *Pauesia rufiabdominalis,* △ *Trioxys acericola,* ▽ *Trioxys quercicola,* ● *Monoctonia pistaciaecola.*

P. exsoletum, Trioxys acalephae, T. brevicornis, T. centaureae, T. complanatus, T. pannonicus. This is a very typical complex of the Eurasian Steppe type areas. It is widely distributed. Due to the cultivation of forest and to the so-called cultivated steppe landscape it has penetrated almost all over the lowland and submountain parts of Europe and it reaches also the semi-desert zone. Some species are also distributed in the Oriental region and in the Far East.

Central Asian Deserts FC. − *Lysiphlebus hispanus* is newly classified as belonging to this complex as it exhibits apparent taxonomical and ecological affinities to *Lysiphlebus desertorum* Starý. It is distributed in semidesert type habitats in the Mediterranean, possibly also in some areas of the Eurasian Steppes (Pannonian lowland).

Other complexes and unclear species. − *Aphidius colemani* and *Pauesia antennata* are members of the Oriental fauna where our present state of knowledge does not yet allow us to distinguish any faunistic complexes of parasites. Formerly *A. colemani* (as *A. transcaspicus*) was regarded as a typical representative of the Mediterranean FC that penetrates to Central Asia (cf. Starý 1968, 1970) but further research on the taxonomy, distribution and host range of this species has indicated its origin to be in the Oriental region; similarly, its occurrence in the Mediterranean is only a part of its distribution range (see Starý 1975).

Lysiphlebus testaceipes is an introduced species that has become established in the mediterranean France and in Corse.

The remaining species, *Aphidius eglanteriae, Praon myzophagum* and *Trioxys auctus* are at present unclear as regards their membership in the particular faunistic complexes.

Relationship to neighbouring areas

A comparison of the particular faunistic complexes in the fauna of the Mediterranean shows that this is a mixture of various faunistic elements, whereas the true Mediterranean species are relatively very rare. Most faunistic connections occur with Central Europe and Central Asia, and a few representatives of the oriental fauna may also be found. On the other hand, no representatives of the Ethiopian fauna have been found. − The present state corresponds fully to the history of floras and faunas which have often found their refugia and spread again to other territories from the Mediterranean area.

Zonation

The aphidiid parasites are primarily distributed in accordance with the main floristic zones irrespective of the altitude above sea level. As it is known, the target zones exhibit the following north − south zonation: forest tundra, coniferous forests, deciduous forests, steppes and forest steppes, deserts and semideserts. The mountain landscapes with their vertical zonation are typical by their interzonal character. The river valleys also exhibit an interzonal character to a considerable extent.

Good examples of zonation and parasite distribution can be presented from the Mediterranean: Italy (STARÝ 1966), Corse (STARÝ, LECLANT, LYON, in press), Iraq (STARÝ and KADDOU 1971) (Photos 1 – 18).

The main interesting differences and the interzonal character of the mountains in this respect are as follows: in the north of Italy, in the Alps (not belonging to the Mediterranean area), the species of the West-Eurasian coniferous forests, of the East Eurasian coniferous forests, European deciduous forests and of Far Eastern deciduous forests, are common; in the lowlands, the species of the Eurasian steppes FC are prevalent. Further to the south, however, the mountain forests of the conifers are composed of the mediterranean species, and some parasites penetrate there in following the same hosts as in the north, of parasitizing new hosts. Some parasite species are already peculiar to this ecosystem. The same is true of the transition of the European deciduous forest into the evergreen sclerophylous forest and shrubs of the Mediterranean. The steppe species of parasites are quite common in the lowlands, on the sea shore, and penetrate to higher altitudes following the rivers and the southern slopes. A similar situation can be found in Corse, but the zonation is much more apparent here as it is concentrated in a relatively small area, besides certain island peculiarities.

On a wider scale, such a zonation can also be presented in Iraq. The oak forests and typical mediterranean landscape can be found in the mountains of Kurdistan, being followed by steppe in the submountains and related belt, and further to the south, semidesert and desert are distributed. A somewhat nontypical situation can be found in the irrigated lands (oases) where the microclimate is different, and these habitats often resemble the interzonal river banks much more than the semidesert neighbourhood; in the irrigated lands we may, for example, find *Lysiphlebus fabarum* as a parasite of aphids on Citrus, besides its occurrence in the herb stratum that is its typical microhabitat. On the other hand, even irrigated lands cannot change the microclimate to such an extent as to allow the occurrence of some species here; for example, *Hyalopterus pruni* is parasitized by *Aphidius colemani* and *Praon volucre* in the mountain orchards and natural communities in the Kurdistan mountains, whereas solely *A. colemani* can occur in the lowland semidesert oasis of Baghdad and southwards; *Macrosiphum rosae*, its parasites in the mountains and their absence in Lower Iraq are a similar example.

Endemics

The analysis of the Mediterranean parasite fauna indicates that the prevalent number of species are also distributed in other areas. A few cases remain unclear: *Pauesia rufiabdominalis* and *P. silana,* parasites of some *Cinara* spp. or *Pinus* in the **41**

Mediterranean. — *Trioxys acericola,* a parasite of *Drepanosiphoniella aceris fugans* in south France (however, the distribution of the host is wider). — *Trioxys quercicola,* a parasite of *Thelaxes suberi* on *Quercus* in Kurdistan mountains. All these species seem to be endemic to the Evergreen sclerophylous and coniferous forests of the Mediterranean.

Island faunas

The parasite fauna of the Canary Islands (MACKAUER 1963), Sicily (STARÝ 1966, TREMBLAY l. c.) and Corse (STARÝ, LECLANT, LYON, in press) has been studied in greater detail, which allows us to summarize the main features as follows: the parasite fauna of the islands of the Mediterranean has the same main features as that in the neighbouring continents but the number of species is generally much less. Whereas there are no significant differences in the species composition in the particular areas of the Mediterranean, it has been found to be reduced both as to the number of parasite species and their hosts in the islands. Several reasons are believed to be responsible for this situation: the primary plant cover of the particular islands, the degree of cultivation and introduction of crops, and, possibly, spread capability of the particular parasite and aphid species. For example, our results obtained in Corse which may be presented as studied in detail have shown that about 50% of species occur there if compared with the mediterranean France; similarly, the number of established records is also less.

Map 3. Distribution range of *Trioxys pannonicus.*

The only record and the following opinion of MACKAUER (1963) on *Trioxys pannonicus* in the Canary Islands could be taken as an opposite to our results. According to MACKAUER (l. c.), the occurrence of this species in the Canaries cannot be connected with any record from Central Europe, the Mediterranean subregion, or from the Middle East. He classified the species as a faunal relic from the warm Tertiary period and as an indication that a former land connection with continental Europe existed. — This opinion has already been opposed by STARÝ (1970, p. 343—4) and the occurrence of *T. pannonicus* in the Canary Islands classified as a common case of distribution of a species belonging to the Eurasian Steppes FC. This opinion has also recently been supported by TREMBLAY's (1972) record of *T. pannonicus* on a small island in the Gulf of Napoli in Italy and by its common occurrence in the Mongolian steppes (STARÝ, unpublished), which allows us to consider *T. pannonicus* as distributed mainly from Central Asia to the western Mediterranean and penetrating to Central and Northern Europe.

Differences in aphid and parasite distribution

Several examples can be presented as follows: (1) Aphid species without parasites in the whole Mediterranean, parasites present in other areas. — The reasons for this have a historical basis connected with changes in the particular floras and faunas, and capability of survival and spread of the aphids and parasites. For example, *Lachnus-*

Map 4. Parasite spectrum of *Hyalopterus pruni* in the Palearctics (original records): ● *Aphidius colemani*, ▽ *Ephedrus persicae*, △ *Ephedrus plagiator*, ■ *Praon volucre*.

43

aphids have parasites in the Far East, but they are unknown in the West Palearctic (cf. STARÝ 1970, 1971). *Callaphis juglandis* (Goetze) is a similar case where parasites are known from Central Asia (cf. STARÝ 1970).

(2) Aphid species with or without parasites in the particular areas of the Mediterranean. — *Cedrobium lapontei* on *Cedrus atlantica* is reported to be parasitized by an aphidiid in North Africa; the aphid has spread and become a serious pest of *Cedrus* in the mediterranean France and Italy, but the parasites have not been found in the latter areas (cf. STARÝ et al. 1971). *Therioaphis trifolii* on alfalfa exhibits different patterns in combination of *Praon exsoletum* and *Trioxys complanatus* in the Mediterranean; these differences are due to different microclimatic requirements of both parasite species (cf. V. D. BOSCH 1957, V. D. BOSCH et al. 1964). — A similar case of such differences are *Macrosiphum rosae* and *Hyalopterus pruni* which exhibit more numerous parasite spectrum in the Kurdistan mountains in the north of Iraq (cf. STARÝ and KADDOU 1971). — *Aphis nerii*: parasites have not been established in the Lower Iraq although the aphid is parasitized in other parts of the Mediterranean; this difference might be due to a peculiar aphid (or parasite?) strain, as parasites were observed to oviposit in the aphids usually in masses but not a single aphid was mummified (cf. STARÝ and KADDOU 1971, STARÝ 1970). — *Aphis spiraecola* and *Toxoptera aurantii* on *Citrus*: they are accidentally introduced pests, parasitized by several indigenous parasite species in the whole Mediterranean. In 1973 — 4 *Lysiphlebus testaceipes* was introduced and has become established in the mediterranean France and Corse and thus the parasite spectrum has become different in this area.

III. BIOLOGICAL PECULIARITIES

Host range

Differences in the host range of a parasite species in dependence on the particular areas of its distribution are rather complicated and need to be analysed in the particular species. Such a classification is, however, beyond the scope of this paper and it seems convenient for the purpose of this paper to present at least some examples as follows:

1. Host range is narrower in the Mediterranean than in Central Europe. — *Praon flavinode* and *Trioxys pallidus*: both species are specific parasites of arboricolous callaphidid aphids in Central Europe. In the Mediterranean, however, they are known

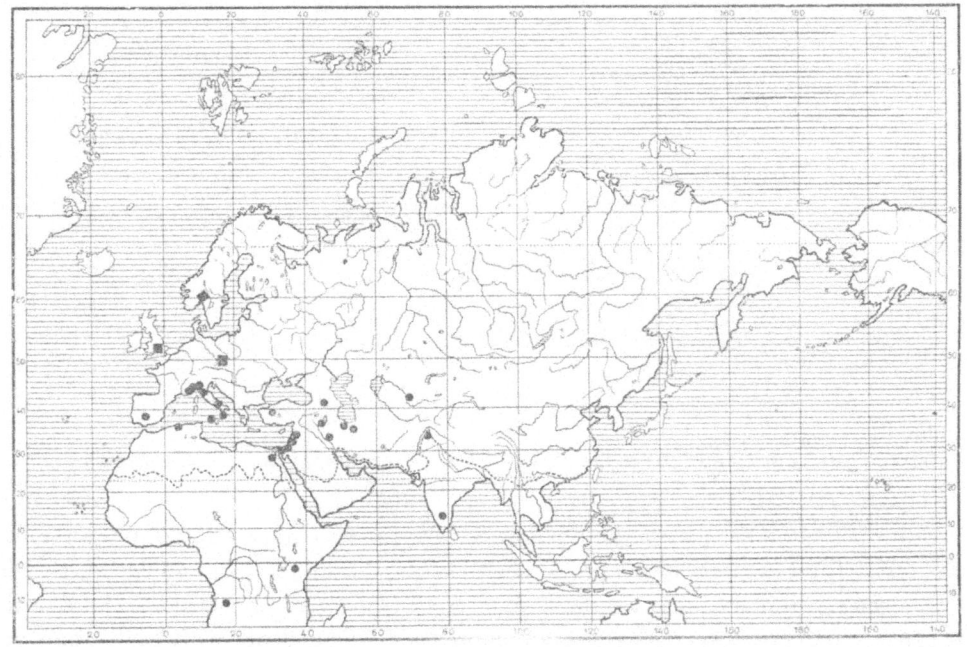

Map 5. Distribution of *Aphidius colemani* in the Palearctics, Oriental and a part of the Ethiopian region. ● natural distribution, ■ purposely introduced (Br. Isles-greenhouses; Czechoslovakia — greenhousess and initial field release experiments). ◆ unclear origin (Norway — greenhouses).

to parasitize a less number of species, although some of them are not distributed in Central Europe.

2. Host range is narrower in the Mediterranean area than in the Oriental region and the tropics. — *Aphidius colemani*: it parasitizes *Hyalopterus pruni* and *Melanaphis donacis* in the West Mediterranean; the same and already some *Aphis* species in the East Mediterranean (Iraq); and in the Oriental, etc., regions the host range is much wider (see Starý 1972, 1975).

3. Host range is wider in the Mediterranean than in Central Europe. — *Lysiphlebus thelaxis*, which is a parasite of *Thelaxes dryophila* in Central Europe, and of *T. dryophila* and *T. suberi* in the Mediterranean. — *Aphidius matricariae*: in Central Europe it is a parasite of *Linosiphon, Capitophorus, Myzus*, etc. aphids but the number of aphid genera and species in the Mediterranean is much larger (cf. Starý 1973 and Starý et al. 1971). — *Monoctonia pistaciaecola*: a parasite of *Pemphigus* species (in galls) in Central Europe, and of *Pemphigus* and *Forda* in the Mediterranean.

Map 6. Distribution and host range of *Monoctonia pistaciaecola*. ● *Forda*, ■ *Pemphigus*.*

A number of differences in the host range of parasites can also be found in the host preference for the particular aphid species. This should be stressed although only a very general indication is available at present (for example, *Aphidius matricariae, A. colemani, A. urticae*). This phenomenon need to be dealt with and taken into consideration in the future research of the Mediterranean; as an adequate study on this problem the review (Chapter I) has been organized correspondingly in the present paper.

Biological races, strains, subspecies

Thelyotokous strains of *Lysiphlebus ambiguus* and *L. fabarum* have been reported to occur in Israel (ROSEN 1967). Although the reasons for the occurrence of thelyotoky in the aphid parasites are not satisfactorily known, we can agree with the classification of Israeli populations as separate strains: *Lysiphlebus ambiguus* is distinctly biparental (arrhenotokous) and *L. fabarum* deuterotokous in Central Europe (cf. STARÝ 1970). Similarly, in Iraq, which is relatively not far from Israel, *L. ambiguus* has been found to be deuterotokous or even arrhenotokous, and *L. fabarum* deuterotokous or arrhenotokous (STARÝ and KADDOU 1971). A number of cases can be found when the same parasite species exhibit certain differences in host spectrum or preference for the particular species, if Central Europe, the Mediterranean and some other areas are taken into consideration. Whether this feature is due to a different population/ strain or environmental factors, is unclear. We are well aware of such cases but a nomenclatorical solution seems to be untimely and perhaps even unjustified.

Another problem are the races or strains adapted to a certain climate. The strain (ecotype) of *Trioxys pallidus* was successfully introduced in 1959 from France into southern California, but it failed to become established in northern or central California; these differences between the sites of collection and introduction are believed to be the cause of the limited success of the French strain (ecotype) as the Iranian strain obtained in 1968 has become established in central California (see V. D. BOSCH et al. 1962, FRAZER and V. D. BOSCH 1973, MESSENGER 1970).

Concerning the subspecies, they were distinguished in three species in the Mediterranean area by MACKAUER (l. c.): (1) *Ephedrus persicae persicae* Froggatt and *E. persicae palaestinensis* Mackauer; the latter was originally described as a separate species (MACKAUER 1959), later classified as a subspecies of *E. persicae* endemic in the eastern Levant (MACKAUER 1963) and later again as a separate species (MACKAUER and STARÝ 1967, MACKAUER 1968). In our opinion, the distinguishing characters of *E. palaestinensis* lie within the variation range of specific characters of *E. persicae* and separation of *E. palaestinensis* on the ground of wing-venation (cf. MACKAUER, l. c.) seems to be unjustified. — (2) *Trioxys angelicae* Haliday. Three subspecies were distinguished by MACKAUER (see MACKAUER and STARÝ 1967, MACKAUER 1968): *T. angelicae angelicae* Haliday — European subregion, *T. angelicae granatensis* Quilis — western Mediterranean, *T. angelicae mediterraneus* Mackauer — eastern Mediterranean; the host range of these subspecies was reported to be the same as in the nominal subspecies. — We have not been able to distinguish these subspecies in the extensive material reared from various aphid hosts in the Mediterranean and thus consider this differentiation of *T. angelicae* into subspecies as unjustified. — (3) *Praon exsoletum* Nees. Two subspecies were distinguished by MACKAUER (see MACKAUER and STARÝ 1967, MACKAUER 1968): *P. exsoletum exsoletum* Nees, reportedly distributed in Europe, exclusive of the Mediterranean area; *P. exsoletum palitans* Muesebeck, distributed in the Mediterranean. — In our opinion, the morphological characters, the host range and distribution of *P. exsoletum* make the separation of the subspecies mentioned above unjustified.

47

Seasonal history and adaptations

Comparatively very little has been known about the seasonal history and adaptations of aphidiid parasites in the Mediterranean.

Obligatory aestival-hibernal diapause has been established in *Monoctonia pistaciaecola* (STARÝ 1966, 1968, 1970), a parasite of *Forda* and *Pemphigus* gall aphids. This is a typical adaptation to the life-cycle of the hosts as the same adaptation of the parasite can be observed in Central Europe, the Mediterranean area and Central Asia, although the distribution of one of the hosts is restricted to the Mediterranean and Central Asia (*Forda* spp. in galls on *Pistacia*; they occur on roots as anholocyclic populations in Central Europe).

On the other hand, obligatory aestival-hibernal diapause has not been determined in *Ephedrus persicae* in the Mediterranean area, although the main features of life-history of the host aphids and the host range of the parasite remain similar both in Central Europe and in the Mediterranean. This phenomenon should be further studied as diapause progeny of *E. persicae* has not been ascertained in West Europe (Netherlands) either, whereas it is known in Central Europe and northern Italy (STARÝ 1966, 1970, 1975).

Aestival diapause was recognized in *Aphidius uzbekistanicus* (STARÝ 1966, as *A. avenae*), a parasite of *Sitobion* spp. on *Gramineae* in Italy; this diapause is believed to be an adaptation for the survival of adverse hot summer conditions, possibly irrespective of the seasonal history of the hosts. It is the diapause of facultative type.

The significance of diapause in parasites for their occurrence in particular areas of the Mediterranean can well be documented by parasites of *Therioaphis trifolii* (*Praon exsoletum* and *Trioxys complanatus*). *T. trifolii* is a holocyclic monoecious species here. Both parasites are specific to *Therioaphis* species. They were introduced into California and detailed studies on this distribution and phenology in California have enabled to understand differences in their distribution patterns in the Mediterranean (cf. V. D. BOSCH 1957, V. D. BOSCH et al. 1964). Both species differ in requirements upon the microclimatic conditions. In some districts of the Mediterranean they can be found occurring together, in others not. Generally, *T. complanatus* is much better adapted (facultative aestival or aestival-hibernal diapause) to hot semidesert climate, whereas *P. exsoletum* prefers milder climate (hibernal diapause), being also distributed in the north of Europe where *T. complanatus* is absent. — Both these parasite adaptations are an apparent example of a direct adaptation/response to the microclimatic conditions.

Ant-attendance

The relationship of the parasites to aphid-attending ants was dealt with by STARÝ (1966, 1969, 1970). According to his observations the ants usually disregard both the mummified aphids and parasite adults if present in an attended colony. There are two exceptions: *Paralipsis enervis* adult is fed by the ants and, besides, the ants mutilate the wings of parasite adults as it is usual in other cases of mutualism. In *Protaphidius wissmannii* the ants nibble the mummies and finally the true parasite cocoon remains, being protected by the ants in a similar way as the live aphids. — The results obtained by STARÝ (l. c.) were supported by ROSEN (1967) who observed the parasites to be very active and eventually exterminating heavily ant-attended colonies of aphids on Citrus.

Interspecific relations

Interspecific relations can be derived from the two basic phenomena: (1) Relative abundance of a parasite in a certain area. It is determined by the requirements of a target species upon the climatic conditions, and by the presence of suitable habitats and hosts (besides the historical factors) in a certain area. This characteristic is even much more apparent if different climatic areas are compared. For example, *Aphidius matricariae* is relatively a rare species in Central Europe, being one of the commonest species in the Mediterranean.

(2) Relative abundance of particular parasite species in the parasite spectrum of a certain aphid species in a certain area. STARÝ (1972) has shown that this abundance is typical of a certain area and can be numerically defined if a sufficient number of parasites is available. — Only field experience records can be presented from the Mediterranean: In Kurdistan mountains (north of Iraq) the parasite spectrum of *Hyalopterus pruni* consists of *A. colemani* and *Praon volucre*, and that of *Macrosiphum rosae* consists of *Aphidius rosae* and *Praon volucre*; on the contrary, in the lowlands of Iraq with semidesert climate *H. pruni* is parasitized solely by *A. colemani*, and *M. rosae* does not have parasites at all (cf. STARÝ and KADDOU 1971). Similarly, parasite spectrum of *Uroleucon* aphids in the Mediterranean consists of *Aphidius funebris*, *Ephedrus niger*, *Praon dorsale* and *Trioxys centaureae*; but solely *A. funebris* was established in Lower Iraq (cf. STARÝ and KADDOU 1971). Similar differences in parasite distribution occur also in Corse in comparison with the mediterranean France (cf. STARÝ, LECLANT, LYON, in press). The parasite spectrum in Corse exhibits less species in this case.

The reasons for certain differences due to specific requirements of *Praon exsoletum* and *Trioxys complanatus*, parasites of *Therioaphis trifolii*, upon microclimatic conditions were documented by V. D. BOSCH et al. (1964; see also above, paragraph on seasonal history).

IV. UTILIZATION IN BIOLOGICAL
AND INTEGRATED CONTROL OF APHIDS

Review of parasites

Indigenous species. — Two species have been utilized in biological control of *Myzus persicae* in glasshouses — *Aphidius matricariae* in Italy (TREMBLAY 1973) and *Diaeretiella rapae* in France (LYON 1968, 1973).

On the other hand, there are a number of species that were ascertained as parasites of key pest aphid species and recommended to be protected in integrated programmes (see below).

Exportations. — The majority of parasite species searched in the Mediterranean area for utilization in biological control in other countries were exported to California, U.S.A. The reasons for this are obvious, as the target pest aphids in California are introduced species, and thus measures were taken to obtain material of indigenous parasites of these pests in a climatically similar area, possibly in the area of their origin. In a lesser degree, parasites were exported to the U.S.S.R. and to Czechoslovakia.

Aphidius colemani Viereck (as *A. transcaspicus* Telenga — exported from Italy and Israel (*Hyalopterus pruni*) to Czechoslovakia in 1964 (STARÝ 1966, 1970). — From southern France (*Melanaphis donacis*) to Czechoslovakia in 1973 (unpublished record). — From Lebanon to California (BIOL. CONTROL INF. BULL. 1967). — From southern France to the U.S.A. (J. J. DREA, USDA Eur. Lab., Sèvres, 1975, in a letter).

Aphidius ervi Haliday: From Lebanon to California in 1962 (COOKE 1963, MACKAUER and FINLAYSON 1967).

Aphidius matricariae Haliday: From France to Maine, U.S.A. in 1957—8 to control potato aphids (SHANDS et al. 1965). — From Israel to California in 1960 (SCHLINGER and MACKAUER 1963).

Diaeretiella rapae M'Intosh: From southern France to the U.S.S.R. for utilization in glasshouses (UŠČEKOV et al. 1972).

Ephedrus persicae Froggatt (as *E. palaestinensis* Mackauer): From Lebanon to California in 1965 (BIOL. CONTROL INF. BULL. 1967).

Lysiphlebus fabarum Marshall: From Lebanon to California in 1965 (MACKAUER and STARÝ 1967).

50 *Lysiphlebus* sp.: From Lebanon to California (BIOL. CONTROL INF. BULL. 1967).

Praon exsoletum Nees: From the Mediterranean area into California in 1955 — 6 (MUESEBECK 1956, V. D. BOSCH 1957).

Praon volucre Haliday: From Israel to California in 1960 (FLESCHNER 1963).

Trioxys angelicae Haliday (as *T. angelicae mediterraneus* Mackauer): From Lebanon to California in 1965 — 6 (BIOL. CONTROL INF. BULL. 1967).

Trioxys complanatus Quilis: From the Mediterranean area to California in 1955 — 6 (MUESEBECK 1956, V. D. BOSCH 1957).

Trioxys pallidus Haliday: From France to California in 1959 (FISCHER et al. 1959, SCHLINGER 1960, V. D. BOSCH et al. 1962).

Introductions. — The number of introduced parasites into the Mediterranean is surprisingly low and may be deduced perhaps from the opinion that the number of indigenous parasites is satisfactory in this ancient area, and, simultaneously due to relatively little interest paid to the utilization of aphidiid parasites in biological control in this area.

Lysiphlebus testaceipes Cresson. This species is widely distributed in the Nearctic America and penetrates far into the Neotropical America. Population of this species was introduced from Cuba to Czechoslovakia and material originating from this laboratory stock was later used for experimental releases in the mediterranean France and in Corse (1973, 1974). The species was initially released in 1973 and became permanently established in the subsequent years. The primary target species were *Aphis spiraecola* and *Toxoptera aurantii*, both introduced pest species on *Citrus*, but the recoveries of the parasite indicate that its wide host range manifests itself also in the Mediterranean and *L. testaceipes* may be expected to control some other pest species (cf. STARÝ, LYON, LECLANT, in press).

Some other parasites have been recommended to be introduced or projects elaborated for parasite search and introduction into the Mediterranean:

Aphidius smithi Sharma et Subba Rao. — This species of Indian origin, a parasite of *Acyrthosiphon pisum,* has become successfully established and good results have been obtained in controlling the pea aphid in the Nearctic America and some other countries. Introduction of *A. smithi* for control of *A. pisum* in the Mediterranean area has recently been proposed by CAMPBELL and MACKAUER (1973). These proposals were discussed by STARÝ (1974) with respect to the indigenous parasites; notes on taxonomy, distinguishing characters, origin, and host range can also be found in the same paper.

Lysiphlebus testaceipes Cresson. — Introduction of this parasite for control of pest aphids on *Citrus* (*Toxoptera aurantii*) and other aphid pests on other plants in the Black Sea coastal area of Gruzia, U.S.S.R. has been proposed by STARÝ (1968). — The recent results obtained in southern France and Corse (cf. STARÝ, LYON, LECLANT, in press) support this project. — On the other hand, the spread capability of *L. testaceipes* indicates that it could become gradually distributed in the whole Mediterranean.

Trioxys nearctaphidis Mackauer. — This is a parasite of *Nearctaphis bakeri* Coven in eastern North America. The aphid has been found as a pest of *Trifolium* in France **51** and the parasite has been recommended for introduction into the Mediterranean (STARÝ, REMAUDIÈRE, LECLANT 1971).

Parasites of Cedrobium laportei Remaudière. — This aphid is associated with *Cedrus atlantica* in North Africa and it has become a serious pest of *Cedrus* trees in the mediterranean France and in Italy. No parasites were found in the latter countries, but mummies were observed in the samples taken in North Africa. Recommendations for search and introduction of this still unknown parasite into France have been given by STARÝ et al. (1971).

Review of the programmes

Apple — Italy: PRINCIPI (1969), PRINCIPI et al. (1967); *Dysaphis plantaginea — Aphidius picipes, Aphis pomi — Trioxys angelicae*; significance in integrated control.

Broad bean (*Vicia faba* L.) — Iraq: AL — AZAWI (1970); *Aphis fabae — Trioxys angelicae, Ephedrus persicae, Lysiphlebus ambiguus, Lysiphlebus fabarum*; proposed timing and selective use of insecticides.

Citrus — Gruzia: STARÝ (1968); *Toxoptera aurantii — Lysiphlebus ambiguus, Lipolexis gracilis*; proposed parasite conservation, and introduction of *Lysiphlebus testaceipes*, multilateral control approach. — Israel: ROSEN (1967); *Toxoptera aurantii — Lysiphlebus ambiguus*; position in integrated programme. — Italy: STARÝ (1964, 1966); *Toxoptera aurantii — Lysiphlebus ambiguus, Lipolexis gracilis, Trioxys angelicae*; proposed augmentation — through reservoirs of economically indifferent alternative host species. — Southern France and Corse: REMAUDIÈRE et al. (1973); indigenous parasites, position in complex integrated control. STARÝ, LYON, LECLANT, in press); indigenous parasites, introduction of *Lysiphlebus testaceipes*; biological control programme. — Cyprus: WOOD (1963); *Toxoptera aurantii — Trioxys angelicae, Ephedrus persicae*; potential value in integrated control.

Peach, apricot — Southern France: REMAUDIÈRE et al. (1973); indigenous parasites, position in a complex integrated control programme. — Italy: STARÝ (1964, 1966); *Hyalopterus pruni — Aphidius colemani, Praon volucre*; proposed augmentation — through reservoirs of economically indifferent aphid species. — Iraq: AL — AZAWI (1970); *Hyalopterus pruni — Aphidius colemani*; proposed timing and selective use of insecticides. AL — RAWY et al. (1969); *Hyalopterus pruni — Aphidius colemani*; proposed timing and selective use of insecticides, on experimental basis.

Tea — Gruzia: STARÝ (1968); *Toxoptera aurantii*; proposed introduction of *Lysiphlebus testaceipes*; relation to *Citrus* groves; multilateral control approach.

Vegetables, ornamentals — Southern France: BILIOTTI and SHARMA (1965), SHARMA (1965); proposed augmentation of *Lysiphlebus fabarum, Aphidius colemani, Diaeretiella rapae* through reservoirs of economically indifferent alternative hosts.

Greenhouses — Southern France: LYON (1968, 1973); *Myzus persicae* — *Diaeretiella rapae*; biological control programme. — Italy: TREMBLAY (1973); *Myzus persicae*, resistant strain — *Aphidius matricariae*; biological control programme.

Ecosystem relations. Multilateral control

Biological and integrated control programmes are usually fully concentrated to a certain key pest species and the target crop ecosystem irrespective of the neighbourhood and existing or possible connections between the neighbouring ecosystems, or only the possible reservoirs of the target pest species in the neighbourhood are taken into consideration. This is undoubtedly the simplest approach but it is rather insufficient for many reasons. An analysis of ecosystem relations, aphid and parasite biology were the main criteria in the new approach to integrated control programme that has been elaborated and defined as *Multilateral control concept* by STARÝ (cf. 1972): according to this concept the aphids (pests) and parasites (natural enemies) exhibit the same or different relations to a target ecosystem and to the particular ecosystems in the neighbourhood. — This viewpoint stresses mainly the role of alternative hosts of parasites in the neighbouring reservoirs. In principle, it has only defined and stressed a common opinion concerning the reservoirs of natural enemies. However, the evaluation of the particular aphid species cannot be generalized, but has to be based on good knowledge of parasite fauna and its host range in a certain area. It is impossible to present here all the cases of ecosystem relations in the Mediterranean area as they differ not only in the particular target ecosystems but to a certain extent also in the same (crop) ecosystem in the particular areas of the Mediterranean. However, several examples have been selected to document and illustrate the Multilateral control approach:

It is quite necessary to know the specific composition of aphids in the target ecosystem and their host range that conditions their possible occurrence in the neighbourhood and the same is needed for the parasites. The particular pest aphid species need to be dealt with at first separately as their parasite spectrum may or may not be identical. The following examples can be analysed:

Alfalfa field. — *Acyrthosiphon pisum* is parasitized by *Aphidius ervi, Praon barbatum*, to a lesser degree by *Aphidius picipes, A. urticae*. Of them, *P. barbatum* is specific to *Acyrthosiphon* aphids and is practically dependent upon the pea aphid population; on the contrary, *A. ervi* exhibits a distinct preference for the pea aphid, but it parasitizes also a number of aphid species (pests) in both cultivated and uncultivated neighbourhood, for example — *Sitobion* spp. on cereals, *Macrosiphum euphorbiae* and *Myzus persicae* on various crops, the same aphid species and *Microlophium evansi, Macrosiphum inexpectatum*, etc. in the waste places and natural environments. *A. picipes* and *A. urticae* prefer the pea aphid to a lesser degree, but their host range and thus relations to the other ecosystems are rather wide. — Another main aphid pest on alfalfa is *Therioaphis trifolii*. Its parasite spectrum is rather specific, **53**

including *Praon exsoletum* and *Trioxys complanatus* which are parasites of *Therioaphis* species. Thus, the parasite spectrum of the pea aphid and of the spotted alfalfa aphid is basically different and the same is true for the relations to the neighbourhood.

Cereal field. — The *Sitobion spp.* aphids are attacked particularly by *Aphidius ervi, A. picipes, A. uzbekistanicus, Ephedrus plagiator, Praon volucre,* to a lesser degree by *A. urticae.* Of them, *A. uzbekistanicus* is specific for *Sitobion* and some other aphids on grasses, whereas the host range of the other parasites (see the host list) is rather broad and enables them to occur and parasitize a number of other aphids in the neighbourhood. In view of integrated control, *A. ervi* is an interesting species as it indicates a close relation between alfalfa (*Acyrthosiphon pisum*) and cereals (*Sitobion*), whereas these crops are quite unrelated if the aphid host range is taken into consideration. This also implies that alfalfa, a perennial crop, and *Acyrthosiphon pisum,* a holocyclic monoecious aphid, represent a perennial reservoir of *A. ervi* in a cultivated area, and this reservoir is important for parasitization of *Sitobion* sp. (monoecious and dioecious species) on annual cereal crops.

Peach, apricot orchards. — If we select *Hyalopterus pruni* from the aphid pests, the situation is as follows (Italy — cf. STARÝ 1964, 1966): *H. pruni* is often heavily parasitized by *Aphidius colemani,* to a lesser degree by *Praon volucre. A. colemani* has *Melanaphis donacis* on *Arundo donax* as an alternative host. Hence, *Arundo donax* groves on waste places are an important reservoir of this valuable species and may be recommended to be even intentionally planted near the orchards (irrigation ditches). — The same opinion has been expressed by BILIOTTI and SHARMA (1965), SHARMA (1965) and concerns also the role of this economically indifferent aphid with respect to the predators (Photo 9, 12).

Citrus groves. — *Aphis spiraecola* and *Toxoptera aurantii,* both accidentally introduced pests, may be selected from the sphid species attacking *Citrus.* They exhibit in principle the same parasite spectrum composed of widely specific parasites (especially *Aphidius matricariae, Lipolexis gracilis, Lysiphlebus ambiguus, L. fabarum, Trioxys angelicae*). All these parasite species can be found as parasites of a number of aphids in the orchard undergrowth and neighbouring windbreaks, whereas the aphid host range is much less (cf. STARÝ 1964, 1966; STARÝ, LYON, LECLANT, in press). The significance of windbreaks with a rich plant species composition (and aphids) is thus apparent: however, *Aphis ruborum* on *Rubus, Aphis arbuti* on *Arbutus, A. rumicis* on *Rumex,* etc. are important here, whereas the common various *Macrosiphoniella* and *Uroleucon* aphid species are quite unimportant as their parasite spectrum is specific, different and does not have any relation to parasites of the *Citrus* pest aphids. — The knowledge of the host range in introduced parasite species is also important. *Lysiphlebus testaceipes* which has recently been introduced for biological control particularly of the citrus pest aphids in the mediterranean France and in Corse has already been found to occur both on citrus (target pests) and on various aphids in the neighbourhood (non-target aphid species which include non-

54

target pests and indifferent species) (see STARÝ, LYON, LECLANT, in press) (Photo 12). — In some cases the Citrus groves (*Toxoptera aurantii*) have close relations to other crops such as tea (also *T. aurantii*); in this case the parasite spectrum is also identical and the same is true for the relation of parasites to the neighbourhood (cf. STARÝ 1968) (Photo 11).

Nerium oleander. — This is a common shrub or tree indigenous to the Mediterranean area, which occurs on the sea shore, river banks, waste places, also in macchia forest, and is also planted as one of the commonest ornamentals in gardens and parks, avenues, etc. The most common aphid associated with oleander is *Aphis nerii*. It is often heavily parasitized by *Lysiphlebus ambiguus, L. fabarum* and *Trioxys angelicae*; all these parasites exhibit a wide host range that includes a number of pest species; hence, the oleander aphid and oleander are an important reservoir of these parasites, whereas *A. nerii* is practically specific to *Nerium* in the Mediterranean. Both *Nerium oleander* and *A. nerii* have become widespread and the aphid also has a wide spectrum of parasites which is different in various parts of the world, according to the presence of members of the particular faunistic complexes of parasites (cf. STARÝ 1968, 1970). Some members of these complexes could be used for introduction to other countries and *A. nerii* can represent one of their important reservoir host species, although the target pest species are different. This is, for example, the case of *Lysiphlebus testaceipes* in France (see STARÝ, LYON, LECLANT, in press). — *A. nerii* and its role in parasite conservation may also be presented as a case of differences that might occur in the particular areas of the Mediterranean: the aphid is commonly parasitized by the same parasite spectrum in the whole Mediterranean except for Iraq where the parasites have not been reared, as they were found to be even incapable of completing their development, although oviposition had been recorded (cf. STARÝ and KADDOU 1971) (Photo 14—16).

Hedera helix, associated with *Aphis hederae*, is of the same importance as the oleander.

The last selected example is to document the necessity of a careful classification of the particular aphids and parasites in the target and neighbouring ecosystems. — We have shown earlier that undergrowth and shrubs (trees in windbreaks) may be regarded as useful reservoirs of parasites that spread from here to the citrus trees and their undergrowth. However, if windbreaks consist of oaks, then there are not any relations, as the oak aphids (*Tuberculoides*, etc.) have a specific parasite spectrum unrelated to that of citrus aphids. The same is true, for example, for the relationship between the alfalfa field and the neighbouring oak hedge (Photo 13).

Several authors have also tried to exemplify the significance of ecosystem relationships in integrated control programmes in the Mediterranean, but their results are not detailed owing to the poor knowledge of the aphid parasites and their host range in this area (cf. BILIOTTI and SHARMA 1965, SHARMA 1965, GRIGOROV 1972).

V. KEY TO THE GENERA AND SPECIES (♀♀)

As mentioned in the introduction, this paper is a review of the parasites of the Mediterranean area as allowed by the present state of our knowledge. The key to the genera and species is based on the corresponding level. However, a number of other genera and species are presumed to be found here in the future and this should always be kept in mind when material from the Mediterranean is identified. If the present key is not found sufficient (sometimes the host records are very useful), recommendations may be given to follow the key to the world genera and subgenera (see STARÝ 1970) and to search for further information pertaining to the particular genera and species mainly in MACKAUER and STARÝ (1967) and MACKAUER (1968).

The names of the genera are printed in bold block letters, those of the subgenera in bold Roman type, and of the species in italics. Two kinds of numbers in the key have also been used: the bold numbers belong to the generic key, the ones in Roman type to the subgenera and species of the particular genera.

Abbreviations and explanations: F — flagellum (F_1, F_2 . . . . flagellar segments 1, 2). Tentorial index = tentorio-ocular line over intertentorial line, relative length (Fig. 43).

1 Median vein developed throughout, separating radial cell 1 from median cell sometimes more or less colourless but distinct in the fore part (Figs. 2, 10)...2
— Median vein effaced frontally or entirely, radial cell 1 and median cell 1 confluent; venation often reduced behind basal vein (Figs. 12, 16, 17, 5, 7, 14, 18) 5
2(1) Both interradial veins developed. (Figs. 2, 4, 5) 4
— Interradial veins effaced (Figs. 9, 10) 3
3(2) Propodeum smooth (Fig. 40). Ovipositor sheaths sparsely haired (Fig. 107) . . . . ***PRAON***

1 F_1 entirely yellow 2
— F_1 dark with yellow to yellowish basal ring, or yellow with at least the apical third dark. 8
2(1) Lateral lobes of mesoscutum pubescent (Figs. 27, 21) 3
— Lateral lobes of mesoscutum with large hairless areas (Figs. 23, 26, 28, 33) 4
3(2) Antennae 18—19-segmented. Wings normal (Fig. 9). Tentorio-ocular line subequal to 1/5 of intertentorial line. F_1 4times as long as wide. Tergite 1 with sharply prominent lateral longitudinal carinae (Fig. 77). . . . .*bicolor*

—	Antennae 21—23-segmented. Wings very ample (Fig. 10). Tentorio—ocular line subequal to 1/3 of intertentorial line. F_1 5 times as long as wide. Tergite 1 coarsely rugose along the sides (Fig. 60). . . . . *grossum*
4(2)	Face densely pubescent (Fig. 44). Antennae 20—21-segmented . . . . *barbatum*
—	Face with central longitudinal narrow hairless area, which is bordered by simple rows of hairs, the area between the rows and orbits with sparse hairs (Figs. 47, 49) 5
5(4)	Antennae 19—20-segmented. Antennae short, reaching hardly to the middle of abdomen . . . . *silvestre*
—	Antennae with another number of segments, long 6
6(5)	Tergite 1 with more or less distinct lateral longitudinal carinae (Figs. 68, 78) 7
—	Tergite 1 more or less rugose along the sides (Fig. 66) . . . . *dorsale*
7(6)	Tergite 1 with sharply prominent lateral longitudinal carinae, the carinae separating a subquadrate prominent area (Fig. 78). Antennae 18—19-segmented . . . . *flavinode*
—	Tergite 1 with more or less prominent lateral longitudinal carinae, which do not separate a prominent area (Fig. 68). Antennae 17—18(16)-segmented . . . . *exsoletum*
8(1)	Tergite 1 distinctly subquadrate, longer than wide at spiracles (cf. Fig. 66). Antennae with various number of segments. Lateral lobes of mesoscutum completely pubescent or with hairless area (Figs. 32, 22) 9
—	Tergite 1 quadrate, as long as wide at spiracles (Fig. 67). Antennae 15—16-segmented. Lateral lobes of mesoscutum with hairless areas (Fig. 39) . . . . *necans*
9(8)	Face with a narrow longitudinal area which is bordered by simple rows of hairs; the area between the rows and orbits practically hairless (Fig. 48). Lateral lobes of mesoscutum with hairless areas (Fig. 22) 10
—	Face with a narrow longitudinal area which is bordered by simple rows of hairs; the area between the rows and orbits with sparse hairs (Fig. 45). Lateral lobes of mesoscutum pubescent or with hairless areas (Fig. 32) 11
10(9)	Antennae 15—16-segmented. Wings subhyaline . . . . *abjectum*
—	Antennae 18—19-segmented. Wings smoky . . . . *absinthii*
11(9)	Antennae 17—19-segmented. Thorax entirely black to dark brown, prothorax sometimes lighter . . . . *volucre*
—	Antennae 15—17-segmented. Lower part of thorax, tergite 1 and apex of abdomen (except dark ovipositor sheaths) yellow . . . . *rosaecola*
Note:	*P. myzophagum* is not included in this key; it is related to *P. volucre*.

—	**Propodeum more or less distinctly areolated (Fig. 42). Ovipositor sheaths densely haired (Fig. 87)** . . . . *AREOPRAON*

(Only 1 species, *A. lepelleyi* is known from the Mediterranean.)

4(2)	**Propodeum regularly areolated, discs of areolae smooth to almost smooth, sometimes slightly sculptured near the carinae (Fig. 30). Ovipositor sheaths with scattered hairs (Fig. 110)** . . . . *EPHEDRUS*

1	Radial abscissa 2 distinctly shorter to subequal to interradial vein 1 (Fig. 6) . . . *persicae*
—	Radial abscissa 2 equal (Fig. 2) or distinctly longer than interradial vein 1 (Fig. 4). 2
2(1)	Radial abscissa 2 equal to interradial vein 1 3
—	Radial abscissa 2 distinctly longer than interradial vein 1 4
3(2)	F_1 long and slender, 1/3 as long as F_2, brown yellow to yellow . . . . *lacertosus*
—	F_1 stout, about 1/6 as long as F_2, black, yellowish at base . . . . *niger*
4(2)	F_1 1/3 as long as F_2. F_1 and a part of F_2 yellowish . . . . *cerasicola*
—	F_1 only slightly longer or equal to F_2; F_1 with only a narrow yellowish base. 5

57

5(4)   Praeapical and apical segments of flagellum thickened, almost fused, forming a club
.... *minor*

—   Praeapical and apical segments of flagellum distinctly separated 6

6(5) F_1 with about 5—7, F_2 — 7, F_3 — 7 rhinaria. Tergite 1 with more or less developed lateral longitudinal carinae and sometimes with more or less distinct central longitudinal carina .... *plagiator*

— F_1 with 1—2 (rarely 0), F_2 — 2, F_3 — 3 rhinaria. Tergite 1 with strongly prominent central and less prominent lateral longitudinal carinae .... *nacheri*

—   Propodeum coarsely and deeply rugose (Fig. 35). Ovipositor sheaths densely pubescent (Fig. 95) .... *LYSEPHEDRUS*

(Only 1 species, *L. validus* is known.)

5(1)   Radial and median cells confluent, distinctly completed by second interradial vein along their external margin (second interradial vein sometimes nearly colourless but distinct) 6

—   Radial and median cells confluent, open, not completed by interradial vein 2 along their external margin 11

6(5)   Confluent radial and median cells distinctly separated on lower margin by fused intermedian and median vein 7

—   Confluent radial and median cells on the lower margin open, the rest of median vein visible only under the second interradial vein.....................10

7(6)   Abdominal segments of normal shape, abdomen lanceolate or rounded 8

—   Abdominal segments beginning with the 4th remarkably tubiform and telescopic (Fig. 57) .... *PROTAPHIDIUS*

(Only 1 species, *P. wissmannii* is known from the West Palearctic.)

8(7)   Ovipositor sheaths slightly curved upwards (Figs. 92, 99, 100, 81)........9

—   Ovipositor sheaths curved downwards, ploughshare-shaped (Figs. 101) .....
.... *MONOCTONUS*

1   Propodeum with large and very wide central areola (Fig. 36)................ *crepidis*

—   Propodeum with narrow central areola (Fig. 31) 2

2(1)   Base and apex of abdomen (including ovipositor sheaths) yellow .... *cerasi*

—   Base and apex of abdomen (including ovipositor sheaths) dark brown to brownish
.... *caricis*

9(8)   Carinae on propodeum forming large wide pentagonal areola (sometimes poorly visible in the longitudinal portion) (Figs. 41, 38, 34) .... *PAUESIA*

1   Ovipositor sheaths wide, stout, only slightly narrowed to the apex and slightly curved upwards (Figs. 108, 92) 2

—   Ovipositor sheaths of other shapes (Figs. 102, 109, 105) 12

2(1)   Flagellar segments (except F_1 and F_2 that are sometimes yellowish or brownish) entirely black 3

—	Flagellar segments beginning with 13—16 yellowish, the apical segment usually darkened .... *infulata*
3(2)	Tergite 1 twice (or almost) as wide at apex as at spiracles, with deep lateral impressions behind spiracular tubercles, strongly dilated to the apex (Fig. 58) 4
—	Tergite 1 only somewhat wider at the apex than at spiracles, with slight lateral impressions behind spiracular tubercles, slightly dilated to the apex (Figs. 59, 54, 53, 55) 8
4(3)	Antennae 22—23-segmented 5
—	Antennae with another number of segments 6
5(4)	Pterostigma brown, yellowish at the base .... *pini*
—	Pterostigma entirely brown .... *abietis*
6(5)	Longitudinal carinae on propodeum absent (Fig. 34). Hind portion of tergite 1 flat, smooth (Fig. 59). Antennae 18—19-segmented .... *silana*
—	Longitudinal carinae on propodeum distinct (Fig. 38). Hind portion of tergite 1 with a central impression and lateral slightly rugose protuberances. Antennae with another number of segments 7
7(6)	Antennae 19—20-segmented .... *silvestris*
––	Antennae 20—21-segmented .... *juniperorum*
8(3)	Mesoscutum strongly raised above prothorax and covering it when viewed from side (Fig. 19) .... *pinicollis*
––	Mesoscutum not covering prothorax as seen from side (Fig. 20) 9
9(8)	Antennae 22-segmented .... *antennata*
—	Antennae with another number of segments 10
10(9)	Antennae 19—20 (18, 21)-segmented 11
—	Antennae 18—19-segmented .... *cupressobii, goidanichi*
11(10)	Tergite 1 more than 3 times as long as wide at spiracles, dilating to the apex and strongly convex in the apical portion (Fig. 55), coarsely longitudinally rugose.... *rufiabdominalis*
—	Tergite 1 2.5 times as long as wide at spiracles, granulate-rugose, almost parallel-sided, slightly convex in the apical portion .... *piceaecollis*
12(1)	Antennae 16—17-segmented. Ovipositor sheaths very slender, only slightly curved upwards, strongly narrowed to the apex (Fig. 105)...................... *unilachni*
—	Antennae with a higher number of segments. Ovipositor sheaths of other shapes (Figs. 102, 109) 13
13(12)	Ovipositor sheaths slender, stout, strongly curved upwards, parallel-sided, bluntly pointed at apex (Fig. 102). Tergite 1 (Fig. 50) .... *picta*
—	Ovipositor sheaths slender, weaker, less curved upwards, narrowed at the base and at apex (Fig. 109). Tergite 1 (Fig. 56) .... *laricis*

—   Carinae on propodeum forming very narrow, small, central areola (Fig. 24) .... *APHIDIUS*

1	Anterolateral area of tergite 1 rugose (Fig. 73) .... *ervi*
—	Anterolateral area of tergite 1 costate (Figs. 75, 76) or costulate (Fig. 74) 2
2(1)	Anterolateral area of tergite 1 costate 3
—	Anterolateral area of tergite 1 costulate 4
3(2)	Tentorial index averaging 0.3. Antennae 17—18-segmented. Ocellar triangle obtuse (Fig. 46) .... *picipes*
—	Tentorial index 0.4—0.5. Antennae 15—16-segmented. Ocellar triangle right to acute (Fig. 46) .... *colemani*
4(2)	Tentorial index 0.6—0.8 5
—	Tentorial index different 6
5(4)	Antennae 18—20(21)-segmented, reaching to the middle of abdomen; apical portion of flagellum not thickened .... *cingulatus*

59

— Antennae 14—15-segmented, reaching to about the end of tergite 1; apical portion of flagellum slightly thickened .... *setiger*

6(4) Tentorial index 0.5—0.6 7

— Tentorial index different 10

7(6) Metacarpus short, about 1/2 shorter than pterostigma 8

— Metacarpus intermediate to long, 1/3 shorter than pterostigma to subequal to its length 9

8(7) Antennae 16—17(18, 15)-segmented. Anterolateral area of tergite 1 with 8 costulae in average (variation 5—12). Temple 1/3—1/4 narrower than transverse eye-diameter, ... *absinthii*

— Antennae 17—18(16)-segmented. Anterolateral area of tergite 1 with 10—12 costulae in average (variation 8—18). Temple 1/6—1/7 narrower than transverse eye-diameter *rosae*

9(7) Anterolateral area of tergite 1 with 4—5 costulae. Antennae 20—21-segmented.................................... *phalangomyzi*

— Anterolateral area of tergite 1 with 10(8—22) costulae. Antennaè 18—19 (17, 20)-segmented .... *funebris*

10(6) Tentorial index 0.4—0.5 11

— Tentorial index 0.3—0.4 13

11(10) Antennae 14—15-segmented .... *hortensis*

— Antennae with a higher number of segments 12

12(11) Antennae 16—17(15, 18)-segmented .... *uzbekistanicus*

— Antennae 18—20(17, 21)-segmented .... *urticae*

13(10) Antennae 12—13-segmented, short, reaching to the end of tergite 1 .... *salicis*

— Antennae 14—18-segmented, longer — reaching to the middle of abdomen 14

14(13) Antennae 14—15-segmented 15

— Antennae 15—16-segmented 16

15(14) Ocellar triangle acute. Wings infumated. F_1 mostly yellow, F_2 yellowish..... *matricariae*

— Ocellar triangle right. Wings hyaline. F_1 with yellowish narrow base, the rest brown .... *eglanteriae*

16(14) Abdomen brown, tergite 1 and the base of tergite 2 yellowish............. *hieraciorum*

— Abdomen yellowish to orange, more or less darker in the central portion. Ovipositor sheaths dark.................................... 17

17(16) Ocellar triangle right. Anterolateral area of tergite 1 with 6—9 costulae. Temple 1/5 narrower than transverse eye-diameter. Clypeus with 7—8 long hairs............ *ribis*

— Ocellar triangle acute. Anterolateral area of tergite 1 with 10—12 costulae. Temple 1/3 narrower than transverse eye-diameter. Clypeus with 12—15 long hairs .... *sonchi*

10(6) Tergite 1 with more or less developed central tubercle only, without central carina or coarse rugosities (Figs. 61, 63, 64). Tentorio-ocular line almost or equal to intertentorial line .... *LYSIPHLEBUS*

1 Flagellar segments distinctly longer than wide. Antennae 12—14-segmented 2

— Flagellar segments distinctly square. Antennae 16—17-segmented (Subg. ***Lysiphlebus***) .... *dissolutus*

2(1) The rest of median and intermedian vein distinct (Figs. 15, 16, 1, 3) Interradial vein distinct, somewhat colourless. Tergite 1 widely triangular (Fig. 62) (Subg. ***Phlebus***) 3

— Only radial vein developed, the rest of median and intermedian vein similarly as interradial vein 2 absent. (Fig. 8). Tergite 1 slender, somewhat dilated to the apex (Fig. 63) (Subg. ***Adialytus***.................................... 6

3(2) Metacarpus distinctly shorter than pterostigma, not reaching wing apex (Figs. 1,15)... 4

— Metacarpus distinctly longer than pterostigma, reaching wing apex (Figs. 3,16)....... 5

4(5) Tergite 1 widely triangular (Fig. 64). Antennae 12-segmented .... *hispanus*

—	Tergite 1 narrowly triangular (Fig. 61). Antennae 13-segmented .... *testaceipes*
5(6)	Lower and apical margin of fore wing with long hairs which are longer than those on the surface (Fig. 3) .... *ambiguus*
—	Lower and apical margins of fore wing with short hairs which are not longer than those on the surface (Fig. 16) .... *fabarum*
6(2)	Apex of abdomen yellowish, lighter than middle segments of abdomen .... *arvicola*
—	Apex of abdomen black brown, not differing from the middle part of abdomen 7
7(6)	F_3—F_4 at least 2.5 times as long as broad, with prevalently adpressed hairs.... *salicaphis*
—	F_3—F_4 twice or less as long as broad, with prevalently semierected hairs .... *thelaxis*

— Tergite 1 with more or less distinct central carina, more or less rugose (Fig. 72). Tentorio-ocular line distinctly shorter than intertentorial line .... **LYSAPHIDUS**

1	Antennae 15(16)-segmented.. *viaticus*
—	Antennae 13-segmented .... *arvensis*

11(5) Radial vein distinctly developed, never pointlike. Legs normal............ 12

— Radial vein pointlike (Fig. 14). Pterostigma large, triangular, strongly sclerotized. Legs strong .... **PARALIPSIS**

(Only 1 species, *P. enervis* is known from the West Palearctic.)

12(11) Ovipositor sheaths curved downwards, terminal abdominal sternite sometimes with 2 prongs 13

— Ovipositor sheaths straight or slightly curved upwards, terminal abdominal sternite without posterior prongs 15

13(12) Terminal abdominal sternite with 2 prongs (Figs. 85, 98) ...... **TRIOXYS**

1	Tergite 1 with primary (= spiracular) and secondary tubercles (Figs. 65, 71, 79), the latter sometimes poorly visible, being almost fused with the primary tubercles (Subg. ***Binodoxys***) (Figs. 69, 79) 2
—	Tergite 1 with primary (= spiracular) tubercles only............................ 6
2(1)	Prongs with 7—8 stout long hairs on the dorsal surface (Fig. 98) .... *centaureae*
—	Prongs with a less number (usually 5) of hairs on the dorsal surface 3
3(2)	Prongs almost straight, only slightly curved at apex (Fig. 114) 4
—	Prongs strongly curved (Figs. 89, 94) 5
4(3)	Distance between primary and secondary tubercles equal or longer than width at spiracles (Fig. 82) .... *angelicae*
—	Distance between primary and secondary tubercles shorter than width at spiracles (Fig. 71) .... *acalephae*
5(3)	Antennae 11-segmented. Primary and secondary tubercles situated in a greater distance from each other (Fig. 80) Apex of prongs with 2 simple bristles (Fig. 89) ..... *heraclei*
—	Antennae 10-segmented. Primary and secondary tubercles situated in a close distance (Fig. 70). Apex of prongs with 1 simple bristle (Fig. 94) .... *brevicornis*
6(1)	Prongs with dilated and strongly differentiated apical portion, with several stout basally dilated bristles (Fig. 97) (Subg. ***Betuloxys***) .... *hortorum*
—	Prongs slightly curved to almost straight, without differentiated apical portion (Figs. 103, 104) (Subg. ***Trioxys***) 7
7(6)	Apex of prongs with bristles of various shape (Figs. 85, 86, 88, 90, 91, 93, 103—4, III.. 8
—	Apex of prongs without bristles (Fig. 115) .... *phyllaphidis*

61

8(7)   Apex of prongs with 2 simple bristles (Fig. 83). Tergite 1 striated in the fore part (Fig. 69) .... *auctus*

—   Apex of prongs with claw-shaped (Figs. 88, 90, 91, 93), uniformly dilated (Figs. 85, 103, 104) or basally dilated (Figs. 86, 111) bristles. Tergite 1 smooth in the fore part 9

9(8)   Apex of prongs with 1 claw-shaped (Figss. 88, 90, 91, 93) bristle 10

—   Apex of prongs with uniformly dilated (Figs. 85, 103, 104) or basally dilated bristles (Figs. 8, 111) 13

10(9)   Prongs straight (Figs. 88, 90) 11

—   Prongs slightly arcuate (Figs. 91, 93)..12

11(10) Ovipositor sheaths broad, slightly more than twice as long as broad, with the constriction on the inner margin near the middle, with very few small scattered hairs near the apex (Fig. 90) .... *complanatus*

—   Ovipositor sheaths narrow, about 3 times as long as broad, with the constriction on the inner margin placed beyond the middle, with small scattered hairs all over the surface except the basal portion (Fig. 88).. *pallidus*

12(10) Prongs with 2 long hairs on dorsal surface (Fig. 91) .... *curvicaudus*

—   Prongs without long hairs on dorsal surface (Fig. 93) .... *quercicola*

13(9)   Apex of prongs with 1 or 4 basally dilated bristles (Figs. 86, 111)..................14

—   Apex of prongs with 2 uniformly dilated bristles (Fig. 85, 103–4) 15

14(13) Apex of prongs with 4 basally dilated bristles (Fig. 111) .... *accricola*

—   Apex of prongs with 1 basally dilated bristle (Fig. 86)...................... *humuli*

15(14) Prongs with 4—7 (or more) long hairs on dorsal surface (Figs. 85, 103) 16

—   Prongs hairless on the dorsal surface except 1 hair before the apex (Fig. 104).... *falcatus*

16(15) Metacarpus shorter than half of pterostigma length. (Fig. 11) .... *pannonicus*

—   Metacarpus longer than half of pterostigma length (Fig. 13) .... *cirsii*

—   Terminal abdominal sternite without prongs 14

14(13) Radial vein longer than 2/3 of its possible length so that pterostigmal cell nearly complete. Ovipositor sheaths slightly curved downwards, their upper part more strongly sclerotized (Fig. 84) .... **LIPOLEXIS**

(Only 1 species, *L. gracilis* is known from the West Palearctic).

—   Radial vein never longer than 2/3 of its possible length; pterostigmal cell distinctly incomplete. Ovipositor sheaths curved downwards and clawed (Fig. 96) .... **MONOCTONIA**

(Only 1 species, *M. pistaciaecola* is known).

15(12) Notaulices entirely effaced. Propodeum with more or less distinct wide central areola (Fig. 29) .... **DIAERETUS**

(Only 1 species, *D. leucopterus* is known).

—   Notaulices at least at the base distinct 16

16(15) Propodeum distinctly areolated, with small central areola. (cf. Fig. 24) ......
.... **DIAERETIELLA**

(Only 1 species, *D. rapae* is known).

—   Propodeum smooth or with 2 divergent carinae in the lower part ....
.... see: **LYSIPHLEBUS**

Figs 1—10

Fore wing: 1 — *Lysiphlebus hispanus*; 2 — *Ephedrus lacertosus*; 3 — *Lysiphlebus ambiguus*; 4 — *Ephedrus plagiator*; 5 — *Lipolexis gracilis*; 6 — *Ephedrus persicae*; 7 — *Diaeretus leucopterus*; 8 — *Lysiphlebus salicaphis*; 9 — *Praon bicolor*; 10 — *Praon grossum*.

Wing-venation: C — Costal vein; Sc — Subcostal vein; Mt — Metacarpus; Pt — Pterostigma; Ptc — Pterostigmal cell; Rc — Radial cell (1, 2, 3); Ir — Interradial vein (1, 2); M — Median vein; Mc — Median cell; Im — Intermedian vein; Bc — Basal cell; B — Basal vein; Cu — Cubital vein; An — Anal vein; Cuc — Cubital cell (1, 2); n — nervulus.

Figs 11—20

Fore wing: 11 — *Trioxys pannonicus*; 12 — *Aphidius rosae*; 13 — *Trioxys cirsii*; 14 — *Paralipsis enervis*; 15 — *Lysiphlebus testaceipes*; 16 — *Lysiphlebus fabarum*; 17 — *Trioxys angelicae*; 18 — *Diaeretiella rapae*; 19 — *Pauesia picta*, mesoscutum, lateral view; 20 — *Pauesia unilachni*, mesoscutum, lateral view.

Figs 21—42

21 — *Praon grossum*, mesoscutum; 22 — *Praon abjectum*, mesoscutum; 23 — *Praon barbatum*, mesoscutum; 24 — *Aphidius rosae*, propodeum; 25 — *Lysiphlebus dissolutus*, propodeum; 26 — *Praon flavinode*, mesoscutum; 27 — *Praon bicolor*, mesoscutum; 28 — *Praon exsoletum*, mesoscutum; 29 — *Diaeretus leucopterus*, propodeum; 30 — *Ephedrus cerasicola*, propodeum; 31 — *Monoctonus caricis*, propodeum; 32 — *Praon volucre*, mesoscutum; 33 — *Praon silvestre*, mesoscutum; 34 — *Pauesia silana*, propodeum; 35 — *Lysephedrus validus*, propodeum; 36 — *Monoctonus crepidis*, propodeum; 37 — *Lysaphidus viaticus*, propodeum; 38 — *Pauesia rufiabdominalis*, propodeum; 39 — *Praon necans*, mesoscutum; 40 — *Praon grossum*, propodeum; 41 — *Pauesia picta*, propodeum; 42 — *Areopraon lepelleyi*, propodeum.

Figs 43—56

43 — *Pauesia* sp., head, frontal view; 44 — *Praon barbatum*, head, frontal view; 45 — *Praon volucre*, head; 46 — Ocellar triangle in *Aphidius*; 47 — *Praon silvestre*, head; 48 — *Praon abjectum*, head; 49 — *Praon dorsale*, head; 50 — *Pauesia picta*, tergite 1; 51 — *Praon necans*, head; 52 — *Praon grossum*, head; 53 — *Pauesia cupressobii*, tergite 1; 54 — *Pauesia goidanichi*, tergite 1; 55 — *Pauesia rufiabdominalis*, tergite 1; 56 — *Pauesia laricis*, tergite 1.

Figs 57—83

57 — *Protaphidius wissmannii*, abdomen. Tergite 1 : 58 — *Pauesia pini*; 59 — *Pauesia silana*; 60 — *Praon grossum*; 61 — *Lysiphlebus testaceipes*; 62 — *Lysiphlebus fabarum*; 63 — *Lysiphlebus salicaphis*; 64 — *Lysiphlebus hispanus*; 65 — *Trioxys angelicae*; 66 — *Praon dorsale*; 67 — *Praon necans*; 68 — *Praon exsoletum*; 69 — *Trioxys auctus*; 70 — *Trioxys brevicornis*; 71 — *Trioxys acalephae*; 72 — *Lysaphidus viaticus*; 73 — *Aphidius ervi*, lateral view; 74 — *Aphidius urticae*, lateral view; 75 — *Aphidius picipes*, lateral view; 76 — *Aphidius colemani*, lateral view; 77 — *Praon bicolor*; 78 — *Praon flavinode*; 79 — *Trioxys quercicola*; 80 — *Trioxys heraclei*; 81 — *Diaeretiella rapae*, genitalia; 82 — *Trioxys angelicae*, tergite 1; 83 — *Trioxys auctus*, genitalia.

Figs 84—96

Genitalia: 84 — *Lipolexis gracilis*; 85 — *Trioxys cirsii*; 86 — *Trioxys humuli*; 87 — *Areopraon lepelleyi*; 88 — *Trioxys pallidus*; 89 — *Trioxys heraclei*; 90 — *Trioxys complanatus*; 91 — *Trioxys curvicaudus*; 92 — *Pauesia silana*; 93 — *Trioxys quercicola*; 94 — *Trioxys brevicornis*; 95 — *Lysephedrus validus*; 96 — *Monoctonia pistaciaecola*.

Figs 97—104

Genitalia: 97 — *Trioxys hortorum*; 98 — *Trioxys centaureae*; 99 — *Aphidius rosae*; 100 — *Lysaphidus arvensis*; 101 — *Monoctonus crepidis*; 102 — *Pauesia picta*; 103 — *Trioxys pannonicus*; 104 — *Trioxys falcatus*.

Figs 105—115

Genitalia: 105 — *Pauesia unilachni*; 106 — *Lysiphlebus fabarum*; 107 — *Praon dorsale*; 108 — *Pauesia abietis*; 109 — *Pauesia laricis*; 110 — *Ephedrus plagiator*; 111 — *Trioxys acericola*; 112 — *Diaeretus leucopterus*; 113 — *Lysaphidus viaticus*; 114 — *Trioxys angelicae*; 115 — *Trioxys phyllaphidis*.

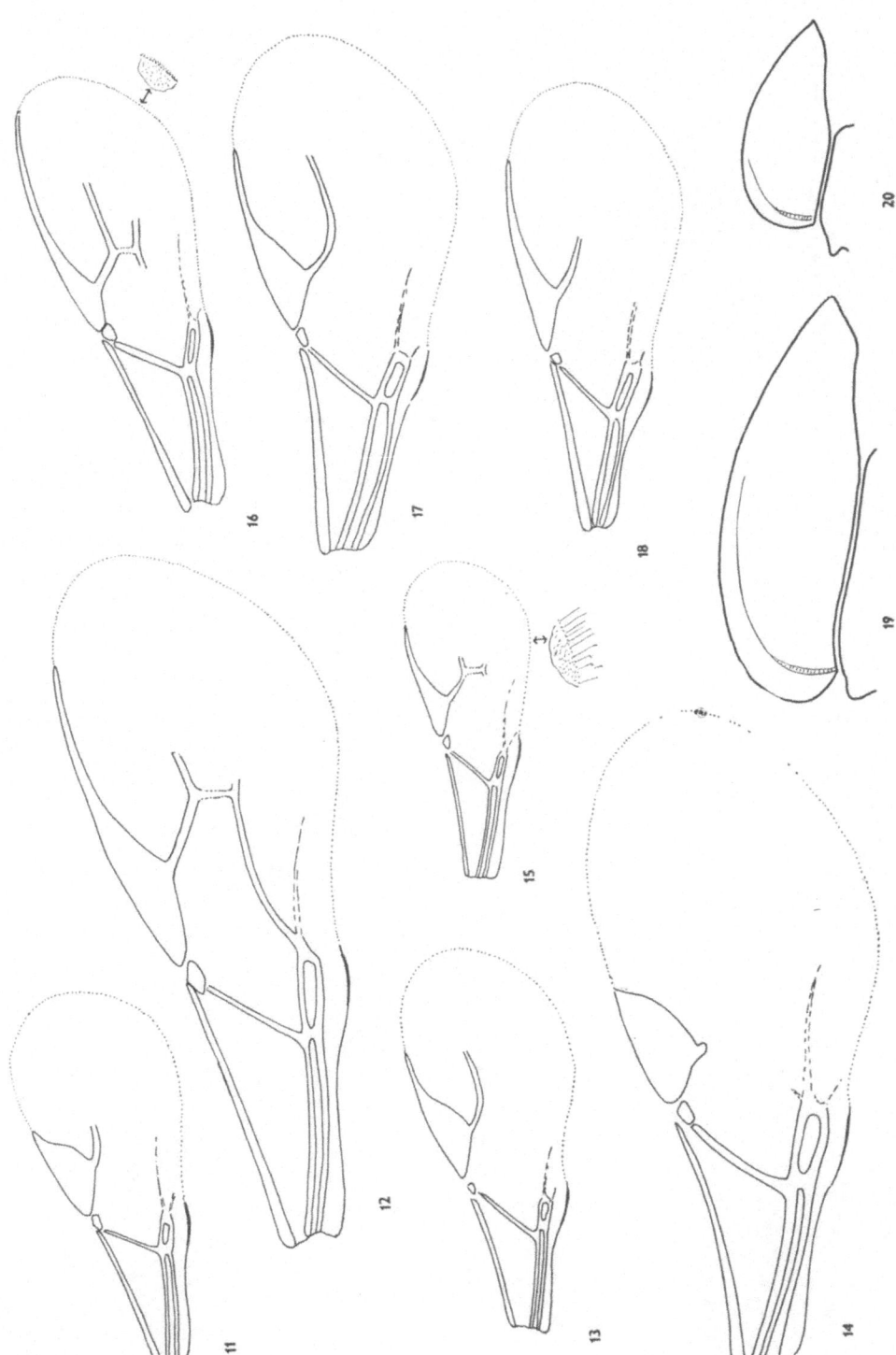

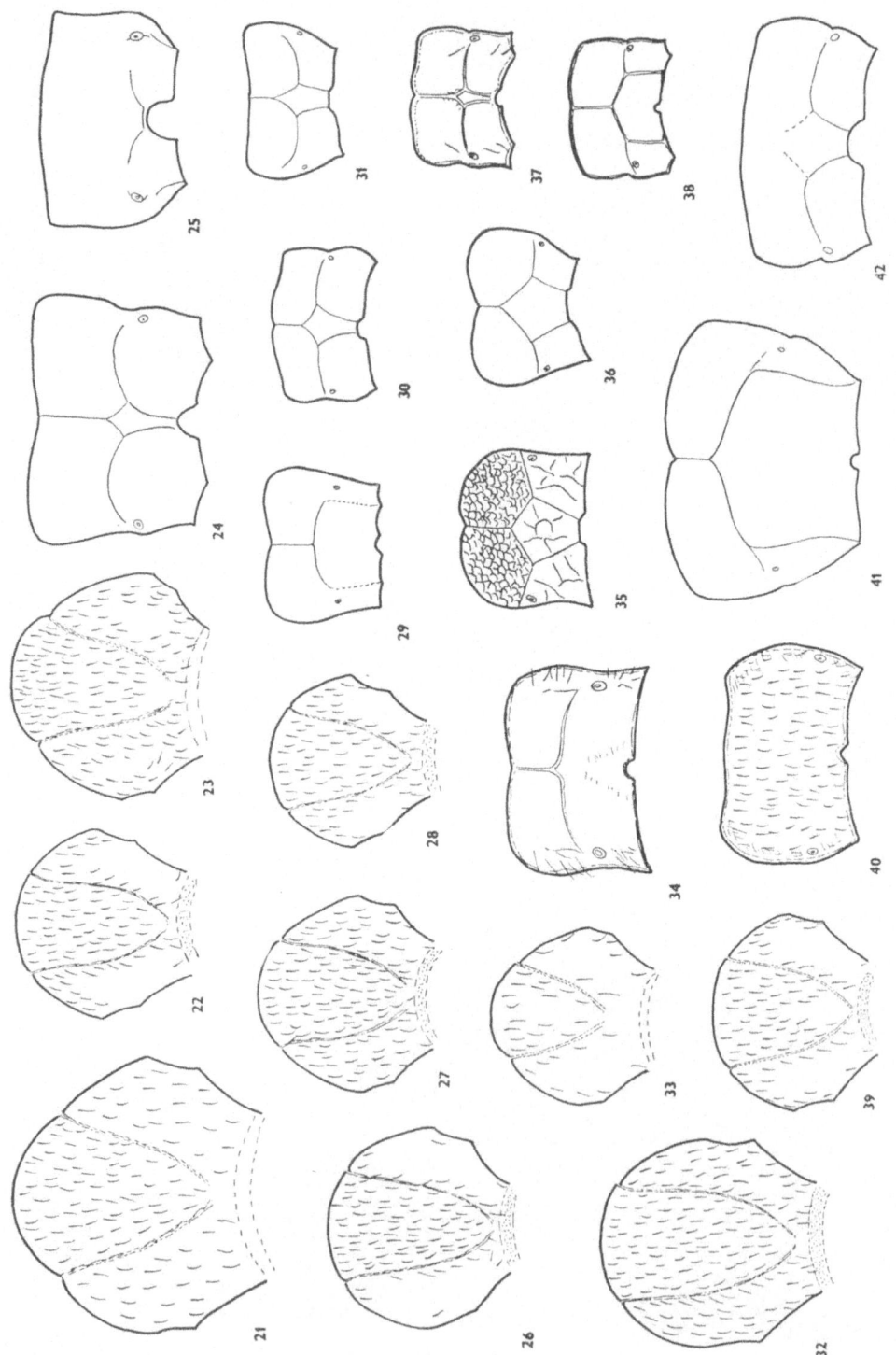

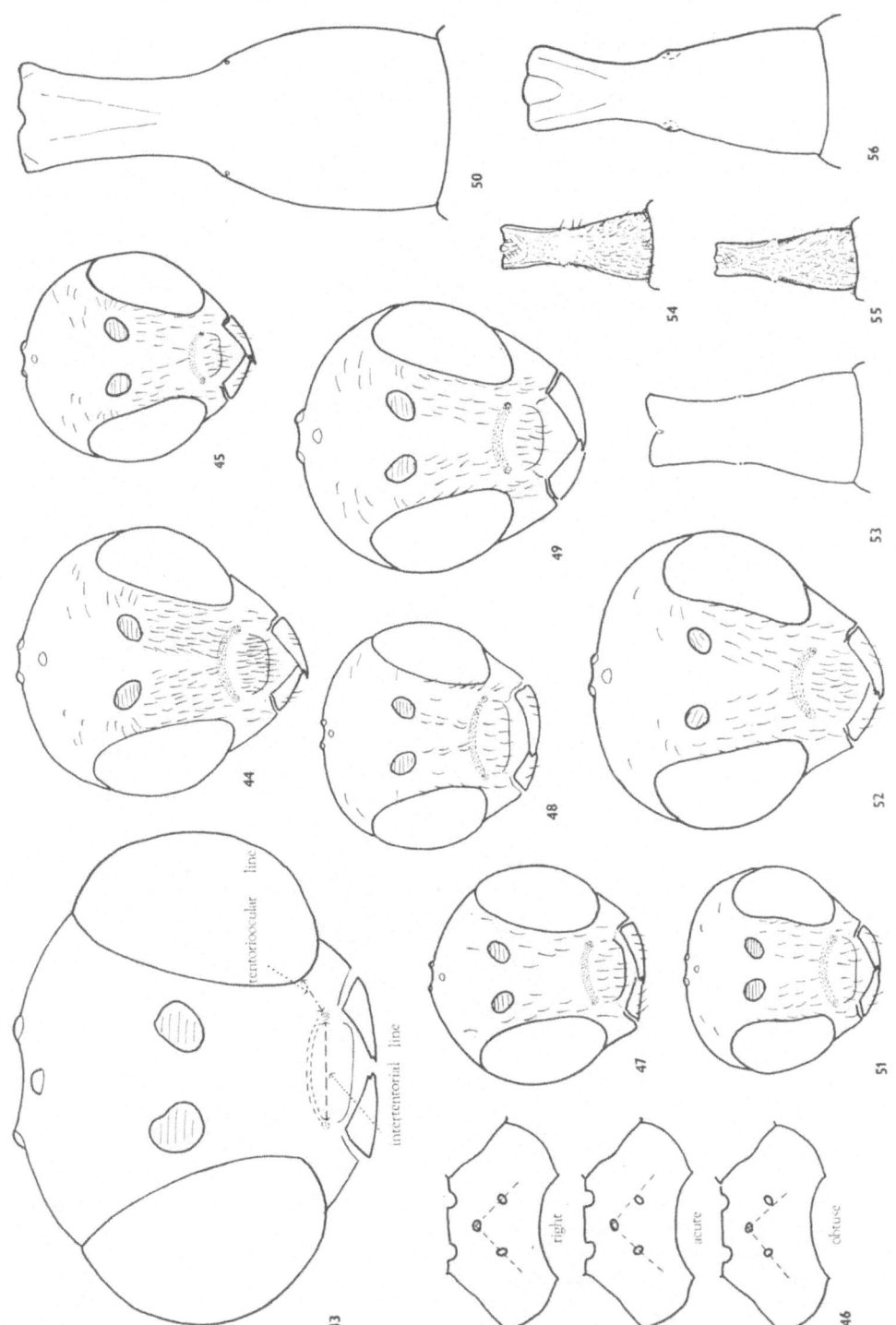

67

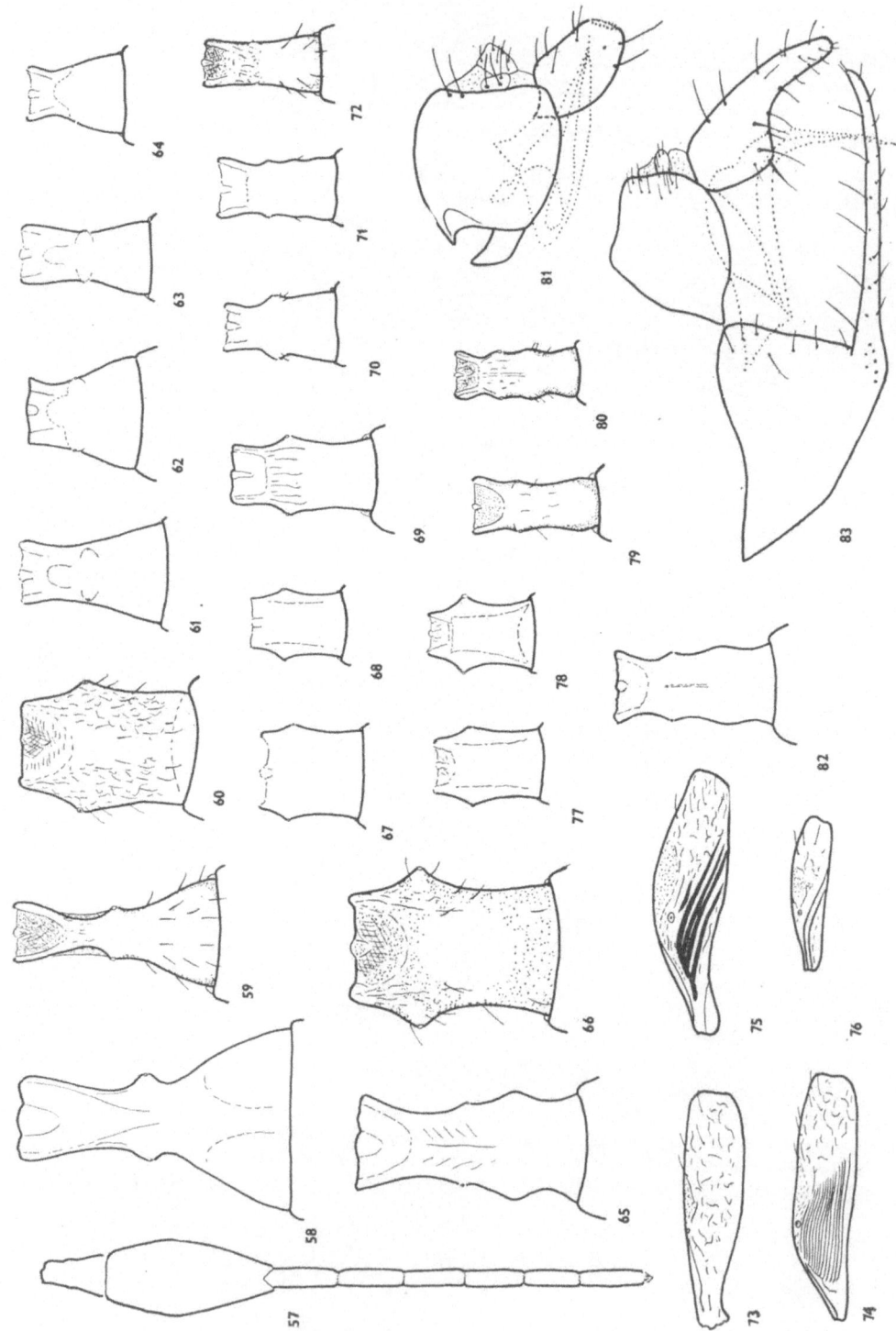

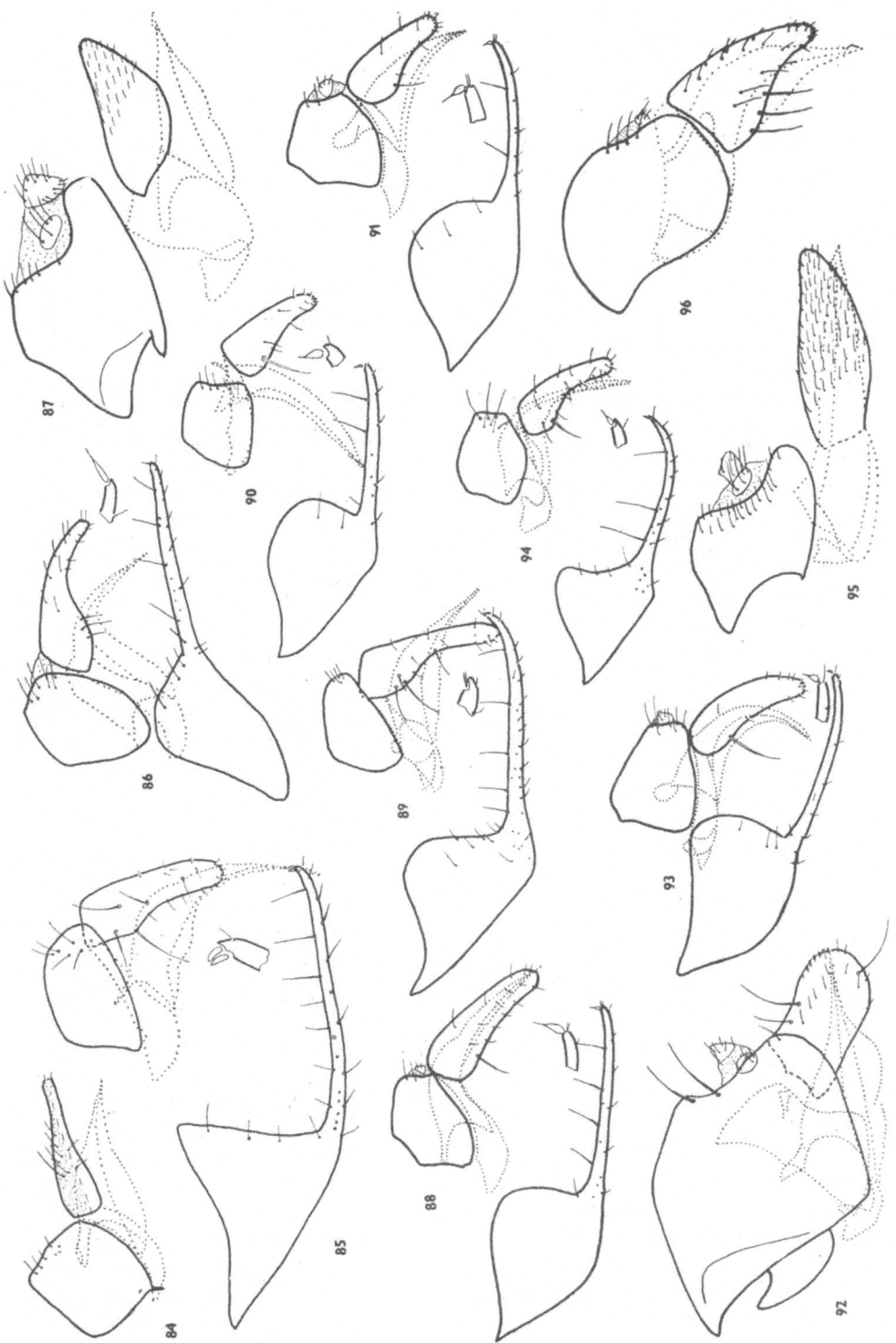

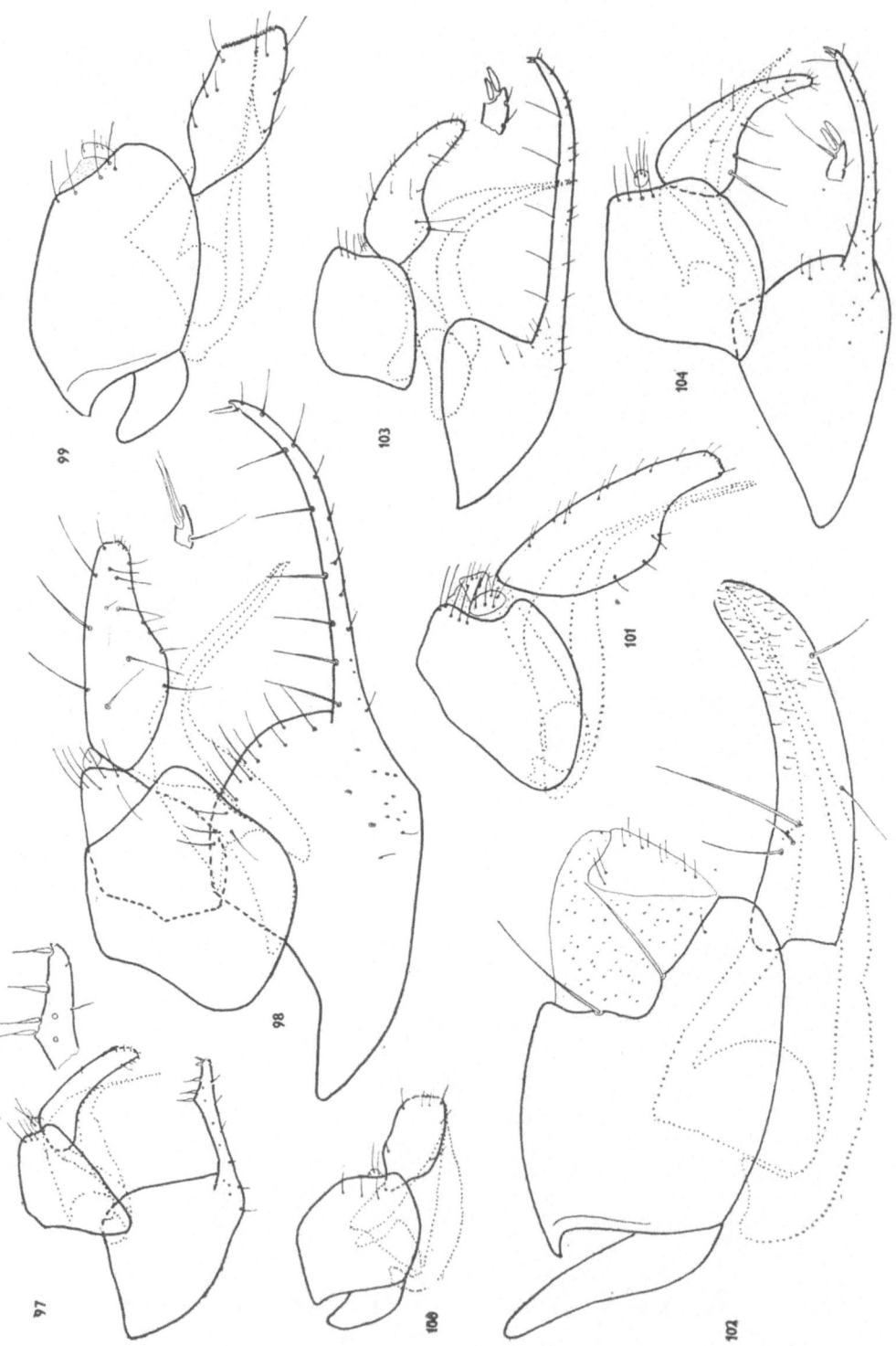

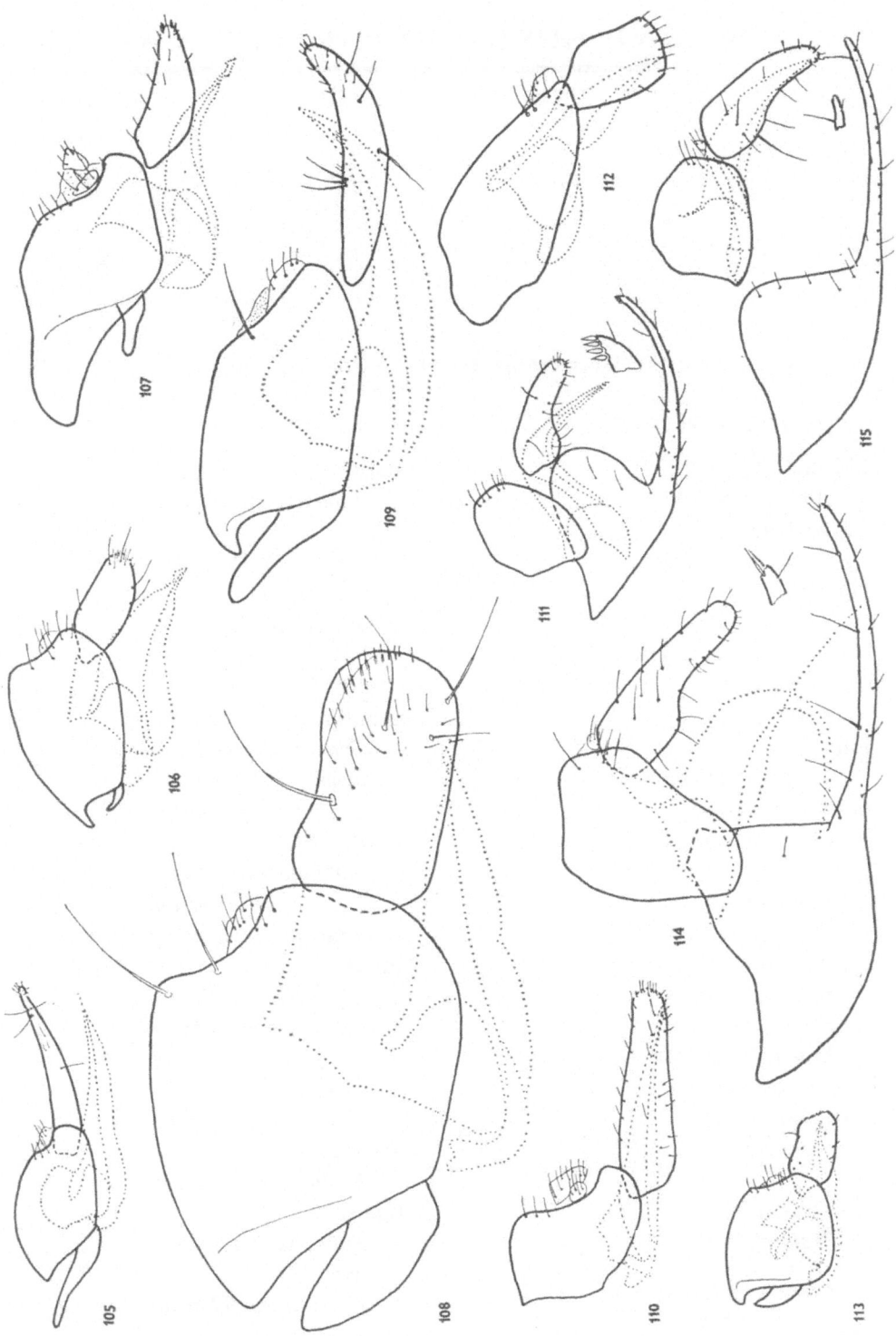

VI. HOST AND PARASITE CATALOGUE

Abbreviations: (!) erroneous record. (?) doubtful record.

ACYRTHOSIPHON

bidentis
Aphidius ervi
euphorbiae
Aphidius urticae
lambersi
Aphidius ervi
Aphidius matricariae
Aphidius sp.
Praon volucre
loti
Aphidius urticae
pelargonii geranii
Praon volucre
pisum
Aphidius ervi
Aphidius picipes
Aphidius urticae
Ephedrus plagiator
Praon barbatum
Praon dorsale (!)
Praon volucre
scariolae
Aphidius sp.
sp.
Aphidius sp.
Praon volucre

AMPHOROPHORA

rubi
Praon grossum

ANOECIA

corni
Lipolexis gracilis

ANURAPHIS

sp.
Ephedrus plagiator

APHIS

acanthi
Diaeretiella rapae
acetosae
Lysiphlebus fabarum
affinis
Aphidius matricariae
Lysiphlebus fabarum
Trioxys acalephae
arbuti
Lipolexis gracilis
Lysiphlebus ambiguus
Lysiphlebus fabarum
Praon volucre
Trioxys angelicae
armata
Lysiphlebus ambiguus
bupleuri
Ephedrus plagiator
chloris
Lysiphlebus fabarum
cisticola
Lipolexis gracilis

Trioxys acalephae
Trioxys angelicae
clematidis
Lysiphlebus ambiguus
Lysiphlebus fabarum
confusa
Lipolexis gracilis
Lysiphlebus fabarum
craccae
Ephedrus plagiator
craccivora
Aphidius matricariae
Ephedrus persicae
Lipolexis gracilis
Lysiphlebus ambiguus
Lysiphlebus fabarum
Praon abjectum
Praon volucre
Trioxys acalephae
Trioxys angelicae
Trioxys sp.
cytisorum
Lysiphlebus fabarum
Trioxys acalephae
dorycnii
Lipolexis gracilis
Lysiphlebus fabarum
durantae
Aphidius sonchi (!)
fabae
Aphidius funebris (!)
Aphidius matricariae
Aphidius sp.
Ephedrus persicae
Ephedrus plagiator
Lipolexis gracilis
Lysiphlebus ambiguus
Lysiphlebus fabarum
Lysiphlebus sp.
Praon abjectum
Praon volucre
Praon sp.
Trioxys acalephae
Trioxys angelicae

farinosa
Aphidius funebris (!)
Lysiphlebus ambiguus
Lysiphlebus fabarum (?)
Trioxys acalephae
frangulae (and cf. frangulae)
Aphidius matricariae
Ephedrus persicae
Lysiphlebus fabarum
Trioxys angelicae
fumanae
Lysiphlebus fabarum
gentianae
Lysiphlebus ambiguus
gossypii
Aphidius matricariae
Aphidius sp.
Ephedrus persicae
Lysiphlebus ambiguus
Lysiphlebus fabarum
Lysiphlebus sp.
Trioxys angelicae
grossulariae
Trioxys acalephae
hederae
Aphidius matricariae
Ephedrus persicae
Lipolexis gracilis
Lysiphlebus fabarum
Trioxys angelicae
idaei
Lysiphlebus fabarum
intybi
Lysiphlebus fabarum
Lysiphlebus testaceipes
lambersi
Aphidius matricariae
Lysiphlebus fabarum
lichtensteini
Ephedrus persicae
Ephedrus plagiator
Trioxys angelicae
medicaginis (& craccivora)
Ephedrus persicae

73

Lysiphlebus ambiguus
Lysiphlebus fabarum
nasturtii
    Aphidius matricariae
    Lysiphlebus ambiguus
    Lysiphlebus fabarum
nerii
    Lysiphlebus ambiguus
    Lysiphlebus fabarum
    Lysiphlebus testaceipes
    Trioxys angelicae
paralios
    Trioxys acalephae
    Trioxys angelicae
parietariae
    Aphidius matricariae
    Ephedrus plagiator
    Lysiphlebus fabarum
    Trioxys acalephae
    Trioxys angelicae
    Trioxys sp.
picridophila
    Lysiphlebus ambiguus
pomi
    Ephedrus plagiator
    Lipolexis gracilis
    Lysiphlebus fabarum
    Praon volucre
    Trioxys angelicae
poterii
    Lysiphlebus fabarum
punicae
    Aphidius colemani
    Lysiphlebus ambiguus
    Lysiphlebus fabarum
    Praon abjectum
    Trioxys angelicae
ruborum
    Aphidius matricariae
    Ephedrus persicae
    Lipolexis gracilis
    Lysiphlebus ambiguus
    Lysiphlebus fabarum
74   *Praon abjectum*

Trioxys acalephae
Trioxys angelicae
rufula
    Lysiphlebus fabarum
rumicis
    Lysephedrus validus (!)
    Lysiphlebus dissolutus (!)
    Lysiphlebus fabarum
    Trioxys angelicae
    Trioxys centaureae (!)
salviae
    Lysiphlebus fabarum
sambuci
    Ephedrus niger (!)
    Praon abjectum
    Trioxys angelicae
sanguisorbae
    Lysiphlebus ambiguus
    Lysiphlebus fabarum
sarothamni
    Lysiphlebus ambiguus
    Lysiphlebus fabarum
sedi
    Lipolexis gracilis
solanella
    Aphidius matricariae
    Lipolexis gracilis
    Lysiphlebus ambiguus
    Lysiphlebus fabarum
    Praon abjectum
    Praon volucre
    Trioxys angelicae
spiraecola
    Aphidius matricariae
    Ephedrus plagiator
    Lysiphlebus fabarum
    Lysiphlebus testaceipes
    Praon abjectum
    Trioxys angelicae
tirucallis
    Lysiphlebus fabarum
umbrella
    Aphidius matricariae
    Diaeretiella rapae

Ephedrus persicae
Lysiphlebus fabarum
Trioxys angelicae
urticata
    Lysiphlebus ambiguus
    Lysiphlebus fabarum
    Trioxys acalephae
vallei
    Lipolexis gracilis
    Lysiphlebus ambiguus
    Trioxys acalephae
verbasci
    Lysiphlebus fabarum
vitalbae
    Lysiphlebus fabarum
viticis
    Lysiphlebus fabarum
    Praon abjectum
    Trioxys angelicae
zizyphi
    Aphidius colemani
    Trioxys angelicae
sp.
    Aphidius colemani
    Aphidius ervi
    Aphidius matricariae
    Aphidius sp.
    Ephedrus persicae
    Ephedrus plagiator
    Lipolexis gracilis
    Lysiphlebus ambiguus
    Lysiphlebus fabarum
    Praon abjectum
    Praon volucre
    Praon sp.
    Trioxys acalephae
    Trioxys angelicae
    Trioxys sp.

AULACORTHUM

solani
    Aphidius picipes
    Praon grossum
    Praon volucre

Praon sp.

BRACHYCAUDUS

amygdalinus
    Diaeretiella rapae
    Ephedrus persicae
    Lipolexis gracilis
    Praon volucre
cardui
    Aphidius matricariae
    Ephedrus persicae
    Lipolexis gracilis
    Lysiphlebus ambiguus
    Lysiphlebus fabarum
helichrysi
    Aphidius matricariae
    Aphidius sp.
    Diaeretiella rapae
    Ephedrus persicae
    Ephedrus plagiator
    Lysiphlebus fabarum
lychnidis
    Praon volucre
mimeuri
    Ephedrus persicae
    Lysiphlebus ambiguus
persicae
    Aphidius absinthii (!)
    Aphidius matricariae
    Ephedrus persicae
    Trioxys angelicae
persicaecola
    Lipolexis gracilis
persicaeniger
    Trioxys angelicae
prunicola
    Lipolexis gracilis
    Lysiphlebus fabarum
prunicola schwartzi
    Lysiphlebus fabarum
rumexicolens
    Diaeretiella rapae
    Ephedrus plagiator
    Lysiphlebus fabarum

spireae
Ephedrus plagiator
tragopogonis
Lysiphlebus ambiguus
Lysiphlebus fabarum
sp.
Aphidius matricariae
Aphidius sp.
Diaeretiella rapae
Ephedrus persicae
Ephedrus plagiator
Lipolexis gracilis
Lysiphlebus ambiguus
Lysiphlebus fabarum
Praon volucre
Praon sp.
Trioxys angelicae
Trioxys sp.

BRACHYUNGUIS

tamaricis
Ephedrus persicae
Lysiphlebus fabarum
tamaricophila
Lysiphlebus fabarum
sp.
Ephedrus persicae
Ephedrus plagiator

BREVICORYNE

brassicae
Diaeretiella rapae

CAPITOPHORUS

carduinus
Aphidius matricariae
Lysiphlebus fabarum
eleagni
Aphidius matricariae
Lysiphlebus fabarum
Trioxys sp.
hippophaes
Aphidius matricariae
Lysiphlebus ambiguus

horni
Aphidius matricariae
inulae
Aphidius matricariae
Ephedrus cerasicola
Ephedrus plagiator
Lysiphlebus fabarum
sp.
Aphidius matricariae
Lysiphlebus fabarum

CAVARIELLA

aegopodii
Aphidius salicis
Lysiphlebus fabarum
Trioxys brevicornis
theobaldi
Trioxys heraclei
sp.
Aphidius salicis
Trioxys heraclei
Trioxys sp.

CEDROBIUM

laportei
unknown aphidiid sp.

CHAETOSIPHON

alpestre
Aphidius eglanteriae
chaetosiphon
Aphidius eglanteriae

CHAITOPHORUS

leucomelas
Lysiphlebus salicaphis
niger
Lysiphlebus salicaphis
tremulae
Lysiphlebus salicaphis
sp.
Lysiphlebus salicaphis
Praon sp.

CHROMAPHIS

juglandicola
Lysiphlebus ambiguus
Trioxys pallidus

CINARA

acutirostris
Pauesia silana
excelsae
Pauesia pini
Pauesia rufiabdominalis
Pauesia silvestris
juniperi
Pauesia cupressobii
Pauesia goidanichi
Pauesia juniperorum
Pauesia sp.
maghrebica
Pauesia piceaecollis
Pauesia silana
nuda
Pauesia unilachni (!)
palestinensis
Pauesia pini (?)
pectinatae
Pauesia infulata
pini
Pauesia picta
Pauesia silana
Pauesia silvestris
pinicola
Pauesia abietis
schimitscheki
Pauesia picta
Pauesia pini
sp.
Pauesia goidanichi
Pauesia juniperorum
Pauesia laricis
Pauesia piceaecollis
Pauesia pini
Pauesia rufiabdominalis
Pauesia silvestris

COLORADOA

bournieri
Lysaphidus arvensis
moralesi
Lysaphidus sp.

CORYLOBIUM

avellanae
Ephedrus plagiator
Praon sp.

CRYPTOMYZUS

ballotae
Aphidius ribis
galeopsidis
Aphidius ribis

CRYPTOSIPHUM

cinae
Lysiphlebus ambiguus

DIURAPHIS

noxius
Diaeretiella rapae

DREPANOSIPHONIELLA

aceris fugans
Trioxys acericola

DYSAPHIS

crataegi
Aphidius sp.
Ephedrus persicae
Ephedrus plagiator
cynarae
Lysiphlebus fabarum
Trioxys angelicae
devecta
Ephedrus persicae
Ephedrus plagiator
Praon sp.
plantaginea
Aphidius matricariae

77

Aphidius picipes
Ephedrus persicae
Ephedrus plagiator
Lysiphlebus fabarum
Trioxys angelicae
pyri
Ephedrus persicae
Ephedrus plagiator
reaumuri
Ephedrus persicae
sorbi
Ephedrus persicae
tulipae
Diaeretiella rapae
sp.
Aphidius sp.
Ephedrus persicae
Lysiphlebus fabarum

ERIOSOMA

lanuginosum
Areopraon lepelleyi
patchae
Areopraon lepelleyi
ulmi
Areopraon lepelleyi

EUCALLIPTERUS

tiliae
Trioxys curvicaudus
Trioxys pallidus

EUCARAZZIA

elegans
Aphidius matricariae
Praon volucre

EULACHNUS

rileyi
Diaeretus leucopterus
Praon bicolor
tuberculostemmata
Diaeretus leucopterus

78 sp.

Diaeretus leucopterus

FORDA

hirsuta
Monoctonia pistaciaecola
follicularia
Monoctonia pistaciaecola
sp.
Aphidius matricariae (?)
Aphidius uzbekistanicus (?)
Monoctonia pistaciaecola

GALIOBIUM

langei
Aphidius matricariae

HAYHURSTIA

atriplicis
Aphidius matricariae
Diaeretiella rapae
Ephedrus nacheri
cadiva
Diaeretiella rapae
Ephedrus nacheri

HOPLOCALLIS

picta
Trioxys pallidus

HYADAPHIS

foeniculi
Aphidius salicis
Ephedrus plagiator
Ephedrus sp.
Trioxys brevicornis
sp.
Ephedrus sp.

HYALOPTERUS

pruni
Aphidius colemani
Aphidius sonchi (!)
Ephedrus persicae
Ephedrus plagiator

Lysiphlebus fabarum
Praon volucre
sp.
Praon volucre (?)
Praon sp.

HYPEROMYZUS

lactucae
Aphidius sonchi
Praon volucre
picridis
Aphidius sonchi
Praon volucre
sp.
Aphidius sonchi
Praon volucre

LINOSIPHON

galiophagus
Praon sp.

LIOSOMAPHIS

berberidis
Aphidius hortensis
Ephedrus plagiator

MACCHIATELLA

sp.
Lipolexis gracilis

MACROSIPHONIELLA

absinthii
Aphidius absinthii
Praon absinthii
artemisiae
Aphidius absinthii
Ephedrus niger
helichrysi
Aphidius absinthii
Ephedrus niger
leucanthemi
Praon absinthii
millefolii
Aphidius absinthii

Aphidius sp.
Ephedrus niger
oblonga
Aphidius phalangomyzi
pulvera
Aphidius absinthii
Ephedrus niger
Lysiphlebus ambiguus
sanborni
Aphidius absinthii
Diaeretiella rapae
Ephedrus niger
Lysiphlebus fabarum
staegeri
Aphidius absinthii
Praon absinthii
tapuskae
Aphidius absinthii
sp.
Aphidius absinthii
Ephedrus niger
Lysiphlebus ambiguus

MACROSIPHUM

doronicicola
Aphidius picipes
Aphidius urticae
Ephedrus plagiator
euphorbiae
Aphidius ervi
Aphidius picipes
Aphidius sp.
Ephedrus plagiator
Praon volucre
holmani
Ephedrus plagiator
inexpectatum
Aphidius ervi
rosae
Aphidius rosae
Ephedrus plagiator
Praon rosaecola
Praon volucre
Praon sp.

saniculae
Ephedrus plagiator
sp.
Aphidius sp.

MASONAPHIS

sp.
Aphidius urticae

MELANAPHIS

donacis
Aphidius colemani
Aphidius uzbekistanicus
Diaeretiella rapae
Lysiphlebus ambiguus
Lysiphlebus fabarum
Praon volucre
pyraria
Aphidius sp.
Ephedrus persicae
Trioxys angelicae

METOPOLOPHIUM

dirhodum
Aphidius uzbekistanicus

MICROLOPHIUM

evansi
Aphidius ervi
Aphidius urticae
Aphidius sp.

MYZOCALLIS

carpini
Trioxys hortorum
Trioxys pallidus
coryli
Trioxys pallidus

MYZUS

cerasi
Aphidius matricariae
Ephedrus persicae
Ephedrus plagiator

Ephedrus sp.
Lipolexis gracilis
Lysiphlebus fabarum
cerasi veronicae
Aphidius matricariae
Lysiphlebus fabarum
ornatus
Aphidius matricariae
persicae
Aphidius ervi
Aphidius matricariae
Aphidius picipes
Aphidius sp.
Diaeretiella rapae
Ephedrus cerasicola
Ephedrus persicae
Lipolexis gracilis
Lysiphlebus fabarum
Praon myzophagum
Praon volucre
Praon sp.
Trioxys angelicae
varians
Aphidius matricariae
Ephedrus plagiator
sp.
Aphidius picipes

NASONOVIA

nigra
Aphidius hieraciorum
Monoctonus crepidis
ribisnigri
Aphidius hieraciorum
Monoctonus crepidis

NEANURAPHIS

rhamni
Lysiphlebus fabarum

PEMPHIGUS

lichtensteini
Lysiphlebus fabarum
sp.

Monoctonia pistaciaecola

PERIPHYLLUS

aceris
Aphidius setiger
bulgaricus
Aphidius setiger
testudinaceus
Aphidius setiger
sp.
Aphidius setiger
Aphidius sp.
Praon silvestre
Trioxys cirsii (!)
Trioxys falcatus

PHORODON

cannabis
Trioxys humuli
humuli
Aphidius matricariae
Ephedrus sp.
Trioxys humuli

PHYLLAPHIS

fagi
Trioxys phyllaphidis

PLEOTRICHOPHORUS

duponti
Lysaphidus viaticus

PROTAPHIS

? terricola
Lysiphlebus fabarum
sp.
Lysiphlebus fabarum
Lysiphlebus hispanus

PTEROCHLOROIDES

persicae
Pauesia antennata

PTEROCOMMA

pilosum
Aphidius cingulatus
populeum
Aphidius cingulatus
sp.
Aphidius cingulatus

RHOPALOSIPHUM

insertum
Ephedrus plagiator
Monoctonus caricis (!)
Monoctonus cerasi
Trioxys auctus
maidis
Aphidius matricariae
Aphidius sonchi (!)
Aphidius uzbekistanicus
Aphidius sp.
Diaeretiella rapae
Ephedrus persicae
Lysiphlebus fabarum
nympheae
Aphidius colemani
padi
Aphidius matricariae
Ephedrus persicae
Lysiphlebus sp.
Praon abjectum
Praon volucre
Trioxys auctus

ROEPKEA

marchali
Ephedrus persicae

SCHIZAPHIS

fritzmuelleri
Aphidius uzbekistanicus
Aphidius sp.
graminum
Aphidius matricariae
Diaeretiella rapae

longicaudata
Diaeretiella rapae

SCHIZOLACHNUS

pineti
Pauesia unilachni
sp.
Pauesia unilachni

SIPHA

maydis
Lysiphlebus arvicola
sp.
Lysiphlebus arvicola

SITOBION

avenae
Aphidius ervi
Aphidius picipes
Aphidius urticae
Aphidius uzbekistanicus
Aphidius sp.
Ephedrus plagiator
Lysiphlebus fabarum
Praon volucre
fragariae
Aphidius ervi
Aphidius urticae
Aphidius uzbekistanicus
Aphidius sp.
Ephedrus plagiator
Praon volucre
sp.
Aphidius ervi
Aphidius urticae
Aphidius uzbekistanicus
Praon rosaecola
Praon volucre

STAEGERIELLA

necopinata
Trioxys brevicornis

STATICOBIUM

limonii
Aphidius absinthii
Praon dorsale
sp.
Praon dorsale

STOMAPHIS

sp.
Protaphidius wissmannii

THELAXES

dryophila
Lysiphlebus thelaxis
suberi
Lysiphlebus thelaxis
Trioxys quercicola
sp.
Lysiphlebus thelaxis

THERIOAPHIS

langloisi
Trioxys complanatus
littoralis
Trioxys complanatus
medicaginis
Praon exsoletum
ononidis
Praon exsoletum
riehmi
Trioxys complanatus
trifolii
Praon exsoletum
Praon sp.
Trioxys complanatus
sp.
Praon exsoletum
Trioxys complanatus

TITANOSIPHON

artemisiae
Trioxys pannonicus

TOXOPTERA

aurantii
Aphidius colemani (?)
Aphidius matricariae
Aphidius picipes
Aphidius sp.
Diaeretiella rapae
Ephedrus persicae
Lipolexis gracilis
Lysiphlebus ambiguus
Lysiphlebus fabarum
Lysiphlebus testaceipes
Praon sp.
Trioxys angelicae
Trioxys sp.
sp.
Lysiphlebus fabarum

TUBERCULOIDES

albosiphonatus
Praon flavinode
Trioxys pallidus
annulatus
Trioxys pallidus
moerickei
Praon flavinode
Trioxys pallidus
sp.
Praon flavinode
Trioxys pallidus

UROLEUCON

achilleae
Ephedrus niger
carthami
Aphidius funebris
Praon dorsale
chondrillae
Aphidius funebris
Praon dorsale

cichorii
Aphidius funebris
Ephedrus niger
Praon dorsale
Praon sp.
hypochoeridis
Ephedrus niger
inulae
Aphidius funebris
Aphidius matricariae (?)
Ephedrus niger
jaceae
Aphidius funebris
Ephedrus niger
Praon dorsale
Praon volucre
Trioxys centaureae
ochropus
Aphidius funebris
picridis
Aphidius funebris
sonchi
Aphidius funebris
Diaeretiella rapae
Ephedrus niger
Praon dorsale
Praon volucre
sp.
Aphidius funebris
Diaeretiella rapae
Ephedrus niger
Lysiphlebus fabarum
Praon dorsale
Praon volucre
Praon sp.
Trioxys centaureae

Other hosts than Aphidoidea (!)
Aphidius sp.
Diaeretiella rapae
Trioxys angelicae

83

SUMMARY

The aphidiid fauna of the Mediterranean area has been reviewed on the ground both of original and published records. The parasite fauna is analysed and divided into the particular faunistic complexes. The relationship of the parasite fauna of the Mediterranean to the neighbouring areas is classified and the peculiarities of zonation, endemics and of island fauna are discussed. Biological peculiarities are summed up with respect to the host range, intraspecific categories, seasonal history and adaptations to the life-cycle of the host, ant-attendance, and interspecific relations. Utilization of parasites in biological and integrated control of aphids is reviewed and the point of view of multilateral control is taken into consideration. The genera and species recognized in the Mediterranean area are keyed. Host and Parasite Catalogue is added.

REFERENCES

ACHVLEDIANI M., 1963: Materials for the study of aphid parasites of east Georgia (in Russ.). Soobšč. Akad Nauk. Gruz. SSR 30 : 781—786.

ACHVLEDIANI M., 1964: K izučeniju sem. Aphidiidae (Hymenoptera) v uslovijach vostočnoj Gruzii. Soobšč. Akad. Nauk Gruz. SSR 33 : 437—440.

AL — AZAWÍ A. F., 1966: Efficiency of aphidophagous insects in Iraq. pp. 277—278. Ecology Aphidoph. Insects, Proc. Symp. Liblice 1965, Academia, Prague.

AL — AZAWÍ A. F., 1970: Some aphid parasites from central and south Iraq with notes on their occurrence. Bull. Iraq nat. Hist. Mus. 4 : 27—31.

AL — AZAWÍ A. F., 1970: Some aphidophagous insects from Iraq with notes on their occurrence. Bull. Iraq nat. Hist. Mus. 4 : 93—104.

ALIEV A. R., 1971: Entomofagi tlej v zapadnych rajonach Azerbajdžana (in Russ.). pp. 10—11. Thesis, Oct. 1971, Biol. metody zašč. plod. i ovošč. kultur. ..... kak osnovy intĕgrirov. sistĕm. Min. S. Ch. SSSR, Kišinĕv, 250 pp.

ALIEV A. R., 1971: Parazity i chiščniki vreditĕlej sada v Azerbajdžanĕ (in Russ.). Proc. 14th Int. Congr. Ent. 2 : 119—120.

AL-RAWY M. A., KADDOU I. K., STARÝ P., 1969; Selectivity of three insecticides used in integrated control of Hyalopterus pruni (Geoffr.) (Homoptera: Aphididae) in Iraq. Bull. Biol. Res. Centre, Baghdad 4 : 13—29.

AL-RAWY M. A., KADDOU I. K., STARÝ P., 1969: Predation of Chrysopa carnea Steph. on mummified aphids and its possible significance in population regulation (Neuroptera, Hymenoptera, Homoptera). Bull. Biol. Res. Centre, Baghdad 4 : 30—40.

ARGYRIOU L. C., 1970: Les aphides nuisibles aux agrumes en Grece et leurs ennemis natureles. Ann. Inst. Phytopath. Benaki N. S. 9 : 114—117.

AVIDOV Z., HARPAZ I., 1969: Plant pests of Israel. Israel Univ. Press., Jerusalem, 549 pp.

AVIDOV Z., KOTTER E., 1966: The pests of safflower Carthamus tinctorius L. in Israel. Scripta Hierosolymitana 18 : 9—26.

BARBAGALLO S., 1974: Osservazioni sugli Afidi (Homoptera, Aphidoidea) del Carciofo (Cynara scolymus L.). Boll. Lab. Ent. Agr. Portici 21: 197—252.

BILIOTTI E., SHARMA M. L., 1965: Les possibilités de lutte biologique contre les pucerons des cultures maraichères et florales de plein air en Provence Maritime. 90e Congr. Soc. Savantes, Nice, 1965 : 563—566.

BIOLOGICAL CONTROL INFORMATION BULLETIN, 1967: Int. Adv. Comm. Biol. Control, Delémont, Switzerland, July 1967, 72 pp.

BODENHEIMER F. S., NEUMARK S., 1955: The Israel pine Matsucoccus (Matsucoccus josephi n. sp.) (Cinara palaestinensis, pp. 63—70). Jerusalem, 122 pp.

BODENHEIMER F. S., SWIRSKI E., 1957: The Aphidoidea of the Middle East. Weizmann Sci. Press, Jerusalem, 378 pp.

VAN DEN BOSCH R., 1957: The spotted alfalfa aphid and its parasites in the Mediterranean region, Middle East, and East Africa. J. econ. Ent. 50 : 352—356.

85

VAN DEN BOSCH R., SCHLINGER E. I., DIETRICK E. J., HALL J. C., PUTTLER B., 1964: Studies on succession, distribution and phenology of imported parasites of Therioaphis trifolii (Monell) in southern California. Ecology 45 : 602—621.

VAN DEN BOSCH R., SCHLINGER E. I., HAGEN K. S., 1962: Initial field informations in California on Trioxys pallidus (Haliday), a recently introduced parasite of the walnut aphid. J. econ. Ent. 55 : 857—862.

CAMPBELL A., MACKAUER M., 1973: Some climatic effects on the spread and abundance of two parasites of the pea aphid in British Columbia (Hymenoptera: Aphidiidae— Homoptera : Aphididae). Z. ang. Ent. 74 : 47—55.

CHALVER R. C., 1973: La familia Aphidiidae (Ins. Him.) en España. Valencia, 312 pp.

COBELLI C., 1897: Gli Imenotteri del Trentino. Notizie preliminari. IV. Publ. Mus. Civ. Rovereto 32 : 1—22.

COBELLI C., 1903: Gli Imenotteri del Trentino. Publ. Mus. Civ. Rovereto 40 : 1—168.

COOKE W. C., 1963: Ecology of the pea aphid in the Blue Mountain area of eastern Washington and Oregon. Tech. Bull. U.S. Dept. Agric. 1287 : 1—48.

DE STEFANI PEREZ T., 1902: Osservazioni biologiche sopra un Braconide aquatico, Giardinaia urinator, e descrizioni di due altri Imenotteri nuovi. Zool. Jahrb., Syst. 15 : 625—634.

EADY R. D., 1969: A new diagnostic character in Aphidius (Hymenoptera : Braconidae) of special significance in species on pea aphid. Proc. R. ent. Soc., London, B, 38 : 165—173.

FAHRINGER J., 1924: Braconidae, Aphidiidae und Serphidae in „Beiträge zur Hymenopterenfauna Dalmatiens, Montenegros und Albaniens" von Dr. F. Maidl. Teil III. Ann. naturhist. Mus. Wien 38 : 98—106.

FISHER T. W., SCHLINGER E. I., VAN DEN BOSCH R., 1959: Imported Trioxys wasp attacks walnut aphid. Diamond Walnut News 41 : 48.

FLESCHNER CH. A., 1963: Releases of recently imported insect parasites and predators in California, 1960—61. Pan-Pacific Ent. 39 : 114—116.

FRAZER B. D., VAN DEN BOSCH R., 1973: Biological control of the walnut aphid in California: the interrelationship of the aphid and its parasite. Env. Entomology 2 : 561—568.

GAPRINDAŠVILI N. K., 1956: Rezultaty izučenija vidovogo sostava i effektivnosti entomofagov kokcid i tlej subtropičeskich kultur Adžarii (in Gruz., Russ. sum.). Trudy Inst. zašč. rast. Gruz. SSR 11 : 173—137.

GOIDANICH A., 1934: Materiali per lo studio degli Imenotteri Braconidi II. Boll. Lab. Ent. Bologna 6 : 209—230.

GOIDANICH A., 1938: Il deperimento primaverile del sorgo zucchrino in Piemonte nei suoi rapporti con gli insetti en in particolare con gli afidi. Boll. Ist. Ent. Bologna 10 : 281—347.

GÓMEZ-MENOR J., 1965: Los Callaphididae de España (Insecta, Homoptera). Bol. R. Soc. Española Hist Nat. (Biol.) 63 : 105—172.

GRAEFFE E., 1908: Beiträge zur Fauna der Braconidae oder Ichneumones adsciti des österr. Küstenlandes and südlichen Krains. Boll. Soc. adriat. Sci. nat. 24 : 137—158.

GRANDI G., 1951: Introduzione allo studio dell'entomologia. Vol. II. Endopterigoti. Ediz. Agric., Bologna, 1332 pp.

GRIGOROV S., 1962: Vklad v entomofaunu Bolgarii (in Bulg.). Rast. Zašč., Sofia 10 : 48—54.

GRIGOROV S., 1967: Contribution to the study on the biology of aphids of Dysaphis genus on apple and their control (in Bulg., Eng., Russ. sum.). Akad Nauk Selskostop. Nauki, Gradina i Logar. Nauka, Sofia 4 : 33—43.

GRIGOROV S., 1972: Interrelationship between certain entomophages and aphids on alfalfa and close-covered wheat crops (in Bulg., Russ. and Engl. sum.). Studies on the biol. control of plant pests, Sofia 1 : 9—29.

HAFEZ M., 1965: Characteristics of the open empty mummies of the cabbage aphid, Brevicoryne brassicae (L.), indicating the identity of the emerged parasites. Agr. Res. Review 43 : 85—88.

HARPAZ I., 1953: Ecology, phenology and taxonomy of the aphids on graminaceous plants in Israel (in Hebrew, with Engl. sum.). Rehovot, 15 pp.

86

HARPAZ I., 1953: Aphids on graminaceous plants (in Hebrew, with Engl. sum.). Ph. D. Thesis, Rehovot, 141 pp.

HARPAZ I., 1955: Bionomics of Therioaphis maculata (Buckton) in Israel. J. econ. Ent. 48 : 668—671.

HASSAN M. S., 1957: Studies on the damage and control of Aphis maidis Fitch in Egypt. Bull. Soc. Entom. Egypte 41 : 213—230.

LISTE D'IDENTIFICATION, 1966: No. 7. Entomophaga 11 : 135—151.

LISTE D'IDENTIFICATION, 1971: No. 8-Comm. tax. entomoph., O.I.L.B., Genève, Suisse, 64 pp.

LIVŠIC I. Z., PETRUŠOVA N. I., 1961: Zaščita plodovogo sada ot vreditělej i bolezněj. Simferopol, Krymizdat, 1961.

LYON J. P., 1968: Remarques préliminaires sur les possibilités d'utilisation pratique d'Hyménoptères parasites pour la lutte contre les pucerons en serre. Ann. Epiphyt. 19 : 113—118.

LYON J. P., 1973: Utilisation des entomophages pour la limitation des populations aphidiennes en serre, pp. 47—9. West Pal. Reg. Sect. Bull. 1973/4 : 73 pp. O.I.L.B.

LYON J. P., 1973: Influence du parasitisme sur les populations de Myzus persicae Sulz. sur les cultures de tomatoes, poivron et aubergines et perspectives de lutte biologique. C. R. 3e Journ. Phytiat. Phytoph. Circum. méditerr. Sassari 1971, Rullière Libeccio, Avignon.

MACKAUER M., 1959: Die europäischen Arten der Gattungen Praon und Areopraon (Hymenoptera : Braconidae, Aphidiinae). Beitr. Ent. 9 : 810—865.

MACKAUER M., 1959: Einige Blattlaus-Schlupfwespen aus Israel (Hym. : Brac., Aphidiidae). Beitr. Ent. 9 : 866—873.

MACKAUER M., 1960: Die europäischen Arten der Gattung Lysiphlebus Förster (Hym. : Braconidae, Aphidiinae). Beitr. Ent. 10 : 582—623.

MACKAUER M., 1961: Neue europäische Blattlaus-Schlupfwespen (Hymenoptera : Aphidiidae). Boll. Lab. Ent. Agr. Portici 19 : 270—290.

MACKAUER M., 1962: Eine neue europäische Lysiphlebus-Art. Mitt. Dtsch. ent Ges. 21 : 12—14.

MACKAUER M. J. P., 1962: Aphid parasites from the Canary Islands. Eos, Madrid 38 : 435—443.

MACKAUER M., 1962: Blattlaus — Schlupfwespen der Sammlung F. P. Müller, Rostock (Hymenoptera : Ichneumonoidea, Aphidiidae). Beitr. Ent. 12 : 631—661.

MACKAUER M., 1963: Bemerkungen zur Systematik, Verbreitung und Wirtsbindung des Ephedrus persicae-Komplexes (Hymenoptera : Aphidiidae). Z. ang. Ent. 52 : 343—354.

MACKAUER M., 1968: Die Aphidiiden (Hymenoptera) Finnlands. Fauna Fennica, Helsingfors 22 : 40 pp.

MACKAUER M., 1968: Aphidiidae. Pars 3. Hymenopterorum Catalogus (n. ed.), edited by FERRIÈRE C. and v. d. VECHT J. Dr. W. Junk N. V., The Hague, 103 pp.

MACKAUER M., CAMPBELL A., 1972: The establishment of three exotic aphid parasites (Hymenoptera : Aphidiidae) in British Columbia. J. entomol. Soc. Brit. Columbia 69 : 54—58.

MACKAUER M., FINLAYSON T., 1967: The hymenopterous parasites (Hymenoptera : Aphidiidae et Aphelinidae) of the pea aphid in Eastern North America. Canad. Ent. 99 : 1051—1082.

MACKAUER M., STARÝ P., 1967: Hym. Ichneumonoidea, World Aphidiidae, in DELUCCHI V. and REUMAUDIÈRE G. (Eds.): Index of entomophagous insects. Le Francois, Paris, 195 pp.

MARSHALL T. A., 1896: Braconides, in ANDRÉ, Species des Hyménoptères d'Europe et d'Algérie. Vol. V. Bouffat Frères, Gray 5 : 635 pp.

MARSHALL T. A., 1897: Braconides, in ANDRÉ, Species des Hyménoptères d'Europe et d'Algérie, Vol. V bis. Bouffat Frères, Gray, 5 (Suppl.): 373 pp.

MARSHALL T. A., 1899: A monograph of British Braconidae. Part VIII. Trans. R. ent. Soc. Lond. 1899 : 1—79.

MARTELLI M., 1911: Notizie sull Aphis brassicae L. e su alcuni suoi parassiti ed iperparassiti. Boll. Lab. Zool. Portici 5 : 40—54.

MENOZZI C., 1932: Il pidocchio (Aphis rumicis L.) dannoso alla bietola da zucchero e la lotta contro di esso. Industr. saccar. Ital. 25 (6). (R.A.E. 1932).

MENOZZI C., 1937: Osservazioni e note entomologiche sulla campagna saccarifera 1936. Industr. saccar. Ital. 30(3) (R.A.E. 1937).

MENTZELOS I. A., 1964: Identification studies of hymenopterous parasites on cotton pests and notes on their biology (in Greek). Rep. Pl. Prot. Agric. Res. Stn., Thessaloniki 2 : 53—62.

MENTZELOS I. A., STATHOPOULOS D. G., SAVVIDIS S. D., 1964: Survey of insects and other pests on crops in Macedonia and Thrace (in Greek). Rep. Pl. Prot. Agric. Res. Stn., Thessaloniki 2 : 71—76.

MESSENGER P. S., 1970: Bioclimatic inputs to biological control and pest management programme, pp. 84—9, in RABB R. L. and GUTHRIE F. E. (Eds.): Concepts of Pest Management, N. C. State Univ., Raleigh.

MIMEUR J. M., 1934: Aphididae du Maroc. Mém. Soc. Sci. nat. phys. Maroc 40 : 1—71.

MIMEUR J. M., 1946: Aleurodidae du Maroc. Bull. Soc. Sci. nat. Maroc 24 : 87—89.

MOKRZECKI S. A., 1913: Report of the Chief Entomologist to the Zemstvo on injurious insects and diseases of plants in the Government of Taurida during the year 1912 (in Russ.). Trudy estěst. istor. Muz., Simferopol 2 : 1—23.

MUESEBECK C. F. W., 1956: Two new parasites of the yellow clover aphid and the spotted alfalfa aphid (Hym., Braconidae). Bull. Brooklyn ent. Soc. 51 : 25—28.

MUESEBECK C. F. W., 1967: Braconidae, pp. 27—60, in KROMBEIN and BURKS, 1967, Hymenoptera of America North of Mexico. IInd Suppl. U.S.D.A. Agric. Monogr. 2 : 584 pp.

PĚLOV V., 1972: Species composition of the entomophages of aphids on fruit trees (in Bulg., with Engl. and Russ. sum.). Studies on the biol. control of plant pests, Sofia 1 : 67—78.

PĚLOV V., 1972: Entomophages of aphids on alfalfa in the Dobrudja (in Bulg., with Engl. and Russ. sum.). Studies on the biol. control of plant pests, Sofia 1 : 49—66.

PĚLOV V., 1972: Studies on the secondary parasites of the genus Asaphes Walk. (Pteromalidae, Hymenoptera) in several regions of Bulgaria and the Soviet Union. Studies on the biol. control of plant pests, Sofia 1 : 29—48.

PIERANTONI U., 1907: Osservazioni sul parassitismo esercitato da un Imenottero (Aphidius aurantii n. sp.) su di un afide degli agrumi (Toxoptera aurantii Fonscol.). Atti Ist. Incorugg. Sci. nat 59 : 3—9.

PRINCIPI M. M., 1969: Recherches en Italie pour l'application de la lutte integrée dans les vergers. Proc. 4th O.I.L.B. Symp. on Integr. Control in orchards, Avignon, pp. 39—43.

PRINCIPI M.M., CASTELLARI P. L., GIUNCHI P., 1967: Observations sur les infestations de pucerons et leurs prédateurs et parasites dans des parcelles traitées avec des produits phytiatriques poly-valents ou selectifs. Entomophaga Mém. Hors. Sér., 3 : 103—107.

QUILIS M. P., 1929: Estudio biológico del icneumónido Aphidius avenae Hal., parásito de los pulgones verdes. Eos, Madrid 5 : 427—459.

QUILIS M. P., 1930: Los parásitos de los pulgones. Dos nuevas especies de Aphidius. Bol. Pat. veg. Ent. Agric., Madrid 4 : 49—64.

QUILIS M. P., 1931: Los parásitos de los pulgones-Notas biológicas sobre les Aphidiidae españoles. An. Inst. nac. 2a Ens. Valencia 20 : 1—36.

QUILIS M. P., 1931: Especies nuevas de Aphidiidae españoles (Hym. Brac.). Eos, Madrid 7 : 25 — 84.

QUILIS M. P., 1932: Tres especies interessantes de Aphidiidae (Hym. Braconidae) de Bologna. Boll. Lab. Ent. Bologna 5 : 49—52.

RADEV R., 1968: Studies on the bioecology of the cotton leaf-aphid Aphis gossypii Glov. (Homoptera, Aphididae) on cotton (in Bulg., Russ. and Engl. sum.). Plant Sci., Sofia 5 : 109—131.

REKAČ V. N., DOBRECOVA T. A., 1933: Cotton aphids in Transcaucasia. Studies on biology and control (in Russ.). Trudy Zakavk. n. issl. Chlop. Inst., Tbilisi 34 : 120.

REMAUDIÈRE G., IPERTI G., LECLANT F., LYON J. P., MICHEL M. F., 1973: Biologie et écologie des aphides et de leurs ennemis naturels; application à la lutte intégrée en vergers. Entomophaga Mém. Hors Sér., 6 : 34 pp.

ROBERTI D., 1969: Nota su afidiini (Hym. Ichneumonoidea) raccolti in Puglia. Entomologica 5 : 101—110.

ROLLI K., 1974: Der Blattlausbefall der Mandelbäume in Tunesien. Gesunde Pfl. 26 : 15—16.

Rondani C., 1847: Osservazioni sopra parecchie species di esapodi aficidi, e sui loro nemici. Nuovi Ann. Sci. Nat. Bologna, Ser. 2, 8 : 337—351, 432—448.

Rondani C., 1848: Osservazioni sopra parecchie species di esapodi aficidi e sui loro nemici. Nuovi Ann. Sci. nat. Bologna, Ser. 2,9 : 5—33.

Rondani C., 1874: Repertorio degli insetti parassiti e delle loro vittime con note ed osservazioni, parte II. Elenco degli insetti dannosi e dei loro parassiti. Boll. Soc. ent. Ital. 6 : 43—68.

Rondani C., 1874: Nuove osservazioni sugli insetti fitofagi e sui loro parassiti fatte nel 1873. Boll. Soc. ent. Ital. 6 : 130—136.

Rondani C., 1876: Repertorio degli insetti parassiti e delle loro vittime. Supplemento alla parte I. Boll. Soc. ent. Ital. 8 : 54—70.

Rondani C., 1877: Vesparia parasita non vel minus observata et descripta. Boll. Soc. ent. Ital. 9 : 166—213.

Rosen D., 1964: Parasites of the Coccoidea, Aphidoidea and Aleurodidea of Citrus in Israel. Thesis. Hebrew Univ. of Jerusalem, 220 pp. (mimeo.).

Rosen D., 1966: Keys for identification of the hymenopterous parasites of scale insects, aphids and aleyrodids on Citrus in Israel. Scripta Hierosolymitana 18 : 43—79.

Rosen D., 1967: On the relationship between ants and parasites of coccids and aphids on citrus. Beitr. Ent. 17 : 281—286.

Rosen D., 1967: Biological and integrated control of citrus pests in Israel. J. econ. Ent. 60 : 1422 – 1427.

Rosen D., 1967: The hymenopterous parasites and hyperparasites of aphids on citrus in Israel. Ann. ent. Soc. Amer. 60 : 394—399.

Rosen D., 1969: The parasites of coccids, aphids and aleyrodids on citrus in Israel: some zoogeographical considerations. Israel Jl. Ent. 4 : 45—53.

Schimitschek E., 1967: Parasitenzuchtergebnisse 1961—1965. Z. ang. Ent. 59 : 64—73.

Schimitschek E., 1944: Forstinsekten der Türkei und ihre Umwelt. Prag, 371 pp.

Schlinger E. I., 1960: The latest import from France. Diamond Walnut News 42 : 9—10.

Schlinger E. I., Mackauer M., 1963: Identity, distribution, and hosts of Aphidius matricariae Haliday, an important parasite of the green peach aphid, Myzus persicae. Ann. ent. Soc. Amer. 56 : 648—653.

Shands W. A., Simpson G. W., Muesebeck C. F. W., Wave H. E., 1965: Parasites of potato-infesting aphids in northeastern Maine. Maine Agric. Exp. Sta, Bull., T 19, Tech. Ser. :77 pp.

Sharma M. L., 1965: Contribution a l'étude de Longiunguis donacis (Pass.) (Aphididae-Homoptera) et des fluctuations de ses populations en Provence Maritime. Théses Fac. Sci., Univ. Paris, A, No. 4150 : 121 pp.

Sifrošvili N., 1971: Rol entomofagov v dinamikě čislennosti listovych tlej na kostočkovych i ich napravlennoje ispolzonavije. Trudy In-ta sadov., vinogr., vinod. Gruz. SSR 1971: 19—20, 47—60.

Starý P., 1961: A revision of the genus Diaeretiella Starý. Acta ent. Mus. Nat. Pragae 34 : 383 — 397.

Starý P., 1962: Faunistic notes on the Aphidiidae of Bulgaria. Acta Faun. ent. Mus. Nat. Pragae 8 : 83—86.

Starý P., 1962: Notes on aphid parasites (Hymenoptera, Aphidiidae) of the southern Crimea (in Russ.). Ent. Obozr. 41 : 875—877.

Starý P., 1962: New aphid parasites of the genus Aphidius from Europe (Hymenoptera, Aphidiidae). Bull. ent. Pologne 32 : 109—122.

Starý P., 1964: Integrated control problems of citrus and peach aphid pests in Italy orchards. Entomophaga 9 : 147—152.

Starý P., 1965: Aphidiid parasites of aphids in the USSR (Hymenoptera: Aphidiidae). Acta Faun. ent. Mus. Nat. Prague 10 : 187—227.

Starý P., 1966: The Aphidiidae of Italy (Hymenoptera: Ichneumonoidea). Boll. Ist. Ent. Bologna 28 : 65—139.

Starý P., 1966: A review of the parasites of aphids associated with Prunus-trees in Czechoslovakia (Hym., Aphidiidae; Hom., Aphidoidea). Acta ent. bohemoslov. 63 : 67—75.

Starý P., 1966: Aphid parasites (Hym., Aphidiidae) and their relationship to aphid attenting ants with respect to biological control. Ins. Sociaux 13 : 185—202.

Starý P., 1967: A review of hymenopterous parasites of citrus pest aphids of the world and biological control projects (Hym., Aphidiidae; Hom., Aphidoidea). Acta ent. bohemoslov. 64 : 37—61.

Starý P., 1968: Biological control of aphids-pests of citrus and tea plantations in the Black Sea coast districts of the USSR-Georgia. Boll. Lab. Ent. Agr. Portici 26 : 227—240.

Starý P., 1968: Diapause in Monoctonia pistaciaecola Starý, a parasite of gall aphids (Hymenoptera : Aphidiidae; Homoptera : Aphidoidea). Boll. Lab. Ent. Agr. Portici 26 : 241—250.

Starý P., 1968: Geographic distribution and faunistic complexes of parasites (Hymenoptera : Aphidiidae) in relation to biological control of aphids (Homoptera : Aphidoidea). Acta Univ. Carolinae, Biologica, Prague, 1967 : 23—89.

Starý P., 1969: Aphids and their parasites associated with oaks in Iraq. Proc. ent. Soc. Wash. 71 : 279—298.

Starý P., 1969: Aphid-ant-parasite relationship in Iraq. Ins. Sociaux 16 : 269—277.

Starý P., 1969: Aphid migration and impact of an indigenous parasite, Aphidius transcaspicus Telenga, on populations of Hyalopterus pruni (Geoffr.) in Iraq (Homoptera : Aphididae and Hymenoptera : Aphidiidae). Bull. Soc. ent. Egypte 53 : 185—198.

Starý P., 1970: Biology of aphid parasites (Hymenoptera : Aphidiidae) with respect to integrated control. Series entomologica 6 : 643 pp. Dr. W. Junk N. V., The Hague.

Starý P., 1971: Parasite spectrum of the Lachnine aphids in the Palearctics: zoogeographical considerations (Hym. Aphidiidae; Hom., Aphidoidea-Lachnidae). Mushi, Fukuoka 44 : 133 – 136.

Starý P., 1972: Host range of parasites and ecosystem relations, a new viewpoint in Multilateral control concept (Hom., Aphididae; Hym., Aphidiidae). Ann. Soc. ent. Fr., N. S. 8 : 351—358.

Starý P., 1972: Relative abundance of parasite species as an area-dependent phenomenon and its possible significance in biological control (Hym., Aphidiidae). Boll. Lab. Ent. Agr. Portici 30 : 19—27.

Starý P., 1973: A review of the Aphidius-species (Hymenoptera, Aphidiidae) of Europe. Annot. zool. bot., Bratislava 84 : 85 pp.

Starý P., 1974: Taxonomy, origin, distribution and host range of Aphidius species (Hym., Aphidiidae) in relation to biological control of the pea aphid in Europe and North America. Z. ang. Ent. 77 : 141—171.

Starý P., 1974: Parasite spectrum (Hym. Aphidiidae) of aphids associated with Galium. Ent. scand. 5 : 73—80.

Starý P., 1975: Parasites (Hym., Aphidiidae) of leaf-curling apple aphids in Czechoslovakia. Acta ent. bohemoslov. 72 : 99—114.

Starý P., 1975: Aphidius colemani Viereck: its taxonomy, distribution and host range (Hym., Aphidiidae). Acta ent. bohemoslov 72 : 156 – 163..

Starý P., Kaddou I. K., 1971: Fauna and distribution of aphid parasites (Hym., Aphidiidae) in Iraq. Acta Faun. ent. Mus. Nat. Pragae 14 : 179—197.

Starý P., Leclant F., Lyon J. P., in press: Aphidiides (Hym.) et Aphides (Hom.) de Corse. Ann. Soc. ent. Fr., N. S.

Starý P., Lyon J. P., Leclant F., in press: Essai de lutte biologique contre les Aphides (Hom.) des Citrus en France méditerranée (Hym., Aphidiidae).

Starý P., Mackauer M., 1971: Trioxys acericola, n. sp. (Hym., Aphidiidae), a parasite of the aphid Drepanosiphoniella from France. Ann. Soc. ent. Fr., N. S. 7 : 885—887.

Starý P., Remaudière G., 1973: Some aphid parasites (Hym., Aphidiidae) from Spain. Entomophaga 18 : 287—290.

Starý P., Remaudière G., Leclant F., 1971 (= Starý et al. 1971): Les Aphidiidae (Hym.) de France et leurs hôtes (Hom., Aphididae). Entomophaga, Mém. Hors Sér. 5 : 72 pp.

STARÝ P., REMAUDIÈRE G., LECLANT F., 1973 (= STARÝ et al. 1973): Nouvelles données sur les Aphidiides de France (Hym.). Ann. Soc. ent. Fr., N. S. 9 : 309—329.

STARÝ P., SCHLINGER E. I., 1967: A revision of the Far East Asian Aphidiidae (Hymenoptera). Series entomologica 3 : 204 pp. Dr. W. Junk N. V., The Hague.

SWIRSKI E., 1957: Fruit-tree aphids in Israel. Thesis, Hebrew Univ. of Jerusalem, 148 pp.

TALHOUK A. S., 1961: Records of entomophagous insects from Lebanon. Entomophaga 6 : 207— 209.

TREMBLAY E., 1961—2: Notulae aphidologicae. I. Notizie su alcuni afidi dannosi. Ann. Fac. Sci. Agr. Univ. Napoli, Portici, Ser. III : 27 : 22 pp.

TREMBLAY E., 1964: Ricerche sugli imenotteri parassiti. I. Studio morfo-biologico sul Lysiphlebus fabarum (Marshall). Boll. Lab. Ent. Agr. Portici 22 : 1—119.

TREMBLAY E., 1965: Gli Insetti entomofagi e la lotta biologica. Le Scienze, Firenze 6 : 310—318.

TREMBLAY E., 1966: Ricerche sugli imenotteri parassiti. II. Osservazioni sull'origine e sul destino dell'involucro embrionale degli Afidiini (Hym. Braconidae : Aphidiinae) e considerazioni sul significato generale delle membrane embrionali. Boll. Lab. Ent. Agr. Portici 24 : 119—166.

TREMBLAY E., 1966: Ricerche sugli imenotteri parassiti III. Osservazioni sulla competizione intraspecifica degli Aphidiinae (Hymenoptera : Braconidae). Boll. Lab. Ent. Agr. Portici 24 : 209—225.

TREMBLAY E., 1967: Ricerche sugli imenotteri parassiti. IV. Notizie su afidiini Italiani (Hymenoptera : Braconidae). Boll. Lab. Ent. Agr. Portici 25 : 59—70.

TREMBLAY E., 1969: Ricerche sugli imenotteri parassiti. VI. Descrizione di una nuova specie del genere Pauesia Quilis Pérez (Hymenoptera : Braconidae : Aphidiinae). Boll. Lab. Ent. Agr. Portici 27 : 153—159.

TREMBLAY E., 1970: Notizie sul complesso Afidi-Afidiini del Pino laricio. Atti VIII. Congr. Naz. Ital. Ent., Firenze 1969 : 119—121.

TREMBLAY E., 1972: Ricerche sugli imenotteri parassiti. IX. Sur un rinvenimento di Trioxys pannonicus Starý (Braconidae : Aphidiinae). Boll. Lab. Ent. Agr. Portici 30 : 131—138.

TREMBLAY E., 1973: The feasibility of using Aphidius matricariae Haliday (Hym., Aphidiidae) against Myzus persicae (Sulz.) (Hom., Aphididae) under glass (in Ital., Engl. sum.). Boll. Lab. Ent. Agr. Portici 30 : 331—350.

TREMBLAY E., CALVERT D., 1971: Embryosystematics in the Aphidiines (Hymenoptera : Braconidae). Boll. Lab. Ent. Agr. Portici 29 : 223—249.

TREMBLAY E., IACCARINO F. M., 1971: Notizie sull'ultrastruttura dei trofociti di Aphidius matricariae Hal. (Hymenoptera : Braconidae). Boll. Lab. Ent. Agr. Portici 29 : 305—313.

TRIGGIANI O., 1973: Contributo alla conoscenza dell'azione svolta dai nemici naturali degli afidi del mandorlo (Amygdalus communis) in agro di Bari. Entomologica 9 : 119—135.

UŠČEKOV A. T., BEGLJAROV G. A., KOZLOVA T. A., 1972: Opyt s perspektivami praktičeskogo ispolzovanija parazita diaeretielly dlja borby v tljami v zaščiščennom gruntě (in Russ.). In: "Biol. metody borby s vredit. ovošč. kultur", Kolos, Moscow, 1972: pp. 43—51.

USŤAN A. K., 1957: Predators and parasites of insects which frequent alfalfa (in Arm.). Izv. biol. s—ch. nauk., AN Arm. SSR, 10(8): 25—29.

VIDANO C., 1959: Analisi morfologica e etologica del ciclo eter. Ogonico di Rhopalosiphum oxyacanthae (Schrank) Börner su Pomoidee e Graminacee. Boll. Zool. Agr. Bachic. 2 : 1—225.

WILKINSON D. S., 1926: Entomological notes. Cyprus agric. J. 21 : 10—12, 47—48.

WILSON F., 1960: A review of the biological control of insects and weeds in Australia and Australian New Guinea. Tech. Comm. Commonw. Inst. Biol. Control 1 : 102 pp.

WOOD B. J., 1963: Imported and indigenous natural enemies of the citrus coccids and aphids in Cyprus, and an assessment of their potential value in integrated control programmes. Entomophaga 8 : 67—82.

ZOULIAMIS N., 1968: Hymenoptera parasites on aphids recorded in Makedonia. Part I. Geoponika, Thessaloniki 14 : 94—95.

91

INDEX OF PARASITE NAMES

Note: Valid names in Roman types.

Records of the Host and Parasite catalogue are not included.

1. *Betula* forest, mountains, Corse. — 2. *Pinus laricio* forest, mountains, Corse. — 3. Sclerophylous forest, mountains, Corse. — 4. Sclerophylous and coniferous forest, submountains, Italy.

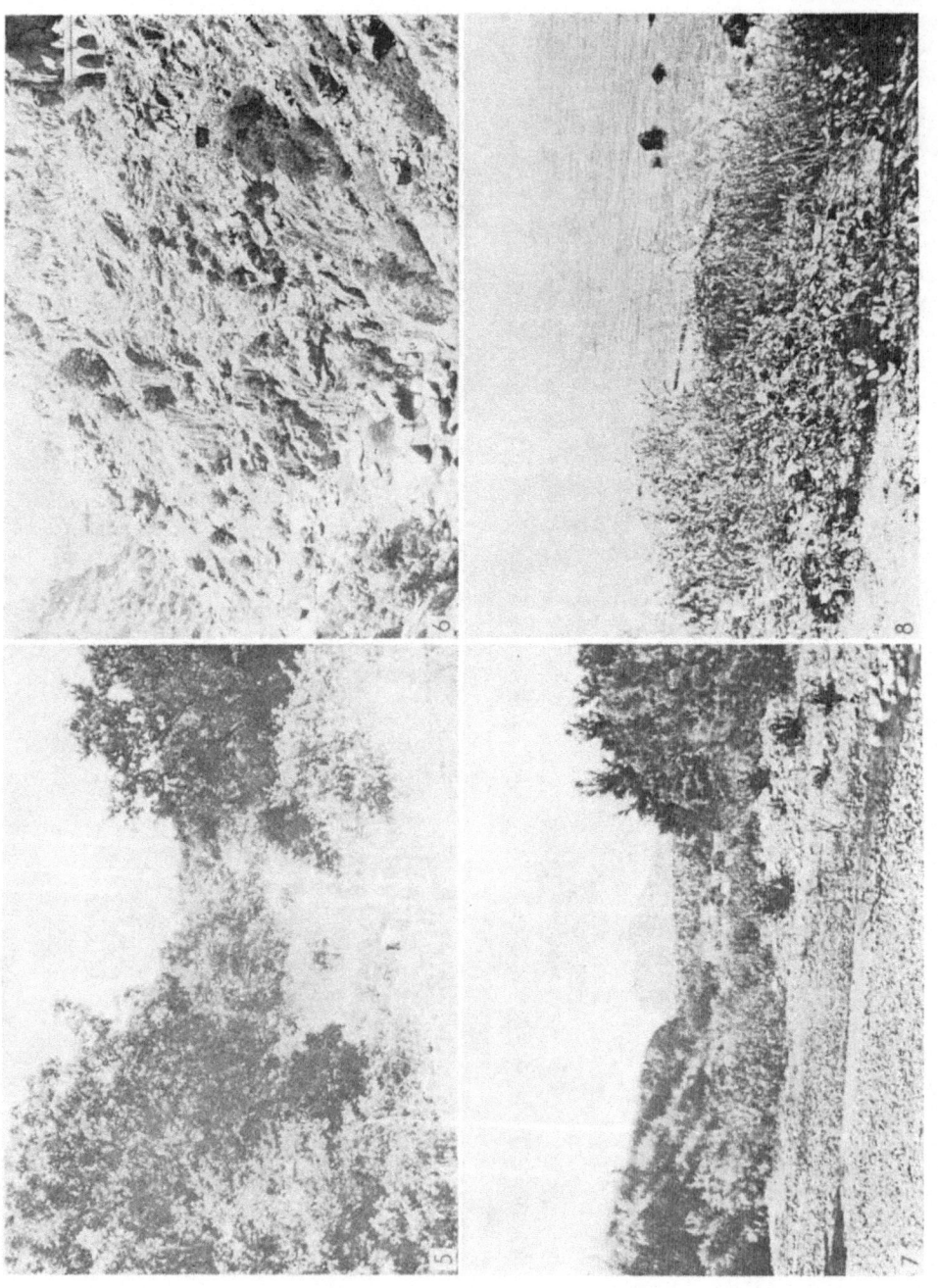

5. Sclerophylous forest, mountains, Corse. — 6. Rocky valley, mountains, Corse. — 7. River valley, submountains, Sicily. — 8. Sea shore, waste place, s. France.

9. *Arundo donax* grove, waste place, Italy. 10. — Grassy habitat, sea shore, Italy. 11. Citrus (right) and tea (left) plantations, Black Sea coastal area, U.S.S.R. — 12. Citrus groves (and windbreaks), modern cultivation system. Citrus Experiment Station, I.N.R.A. San Giuliano. Corse.

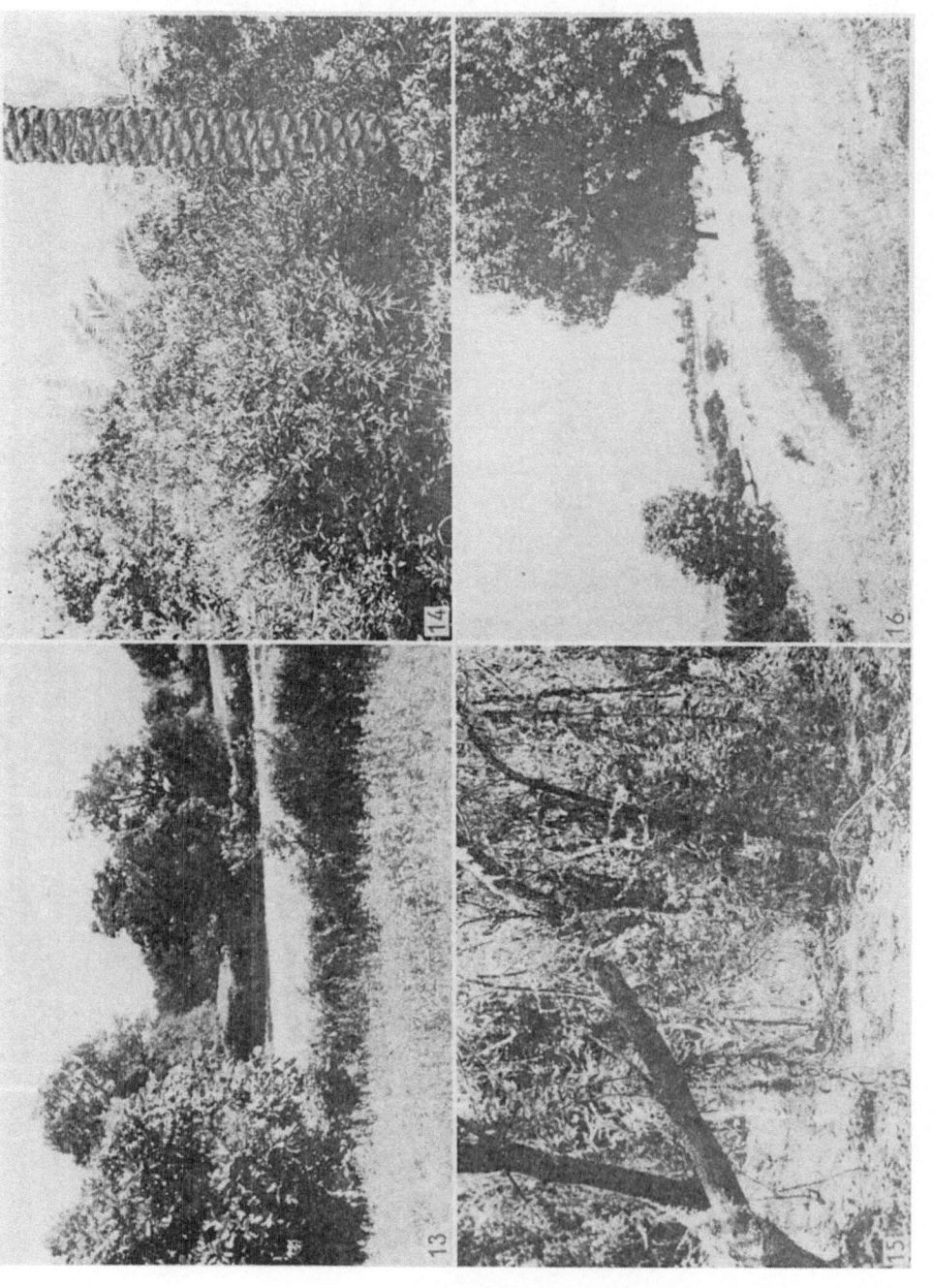

13. Sclerophylous forest and cultivated (alfalfa) plots, s. Yugoslavia. — 14. Irrigated gardens, Baghdad, Iraq. — 15. Lowland riverain forest, *Populus euphratica*, Lower Iraq. — 16. Open oak (*Quercus* spp.) woodland and riverain forest (left-*Populus nigra*), mountains, Kurdistan, Iraq.

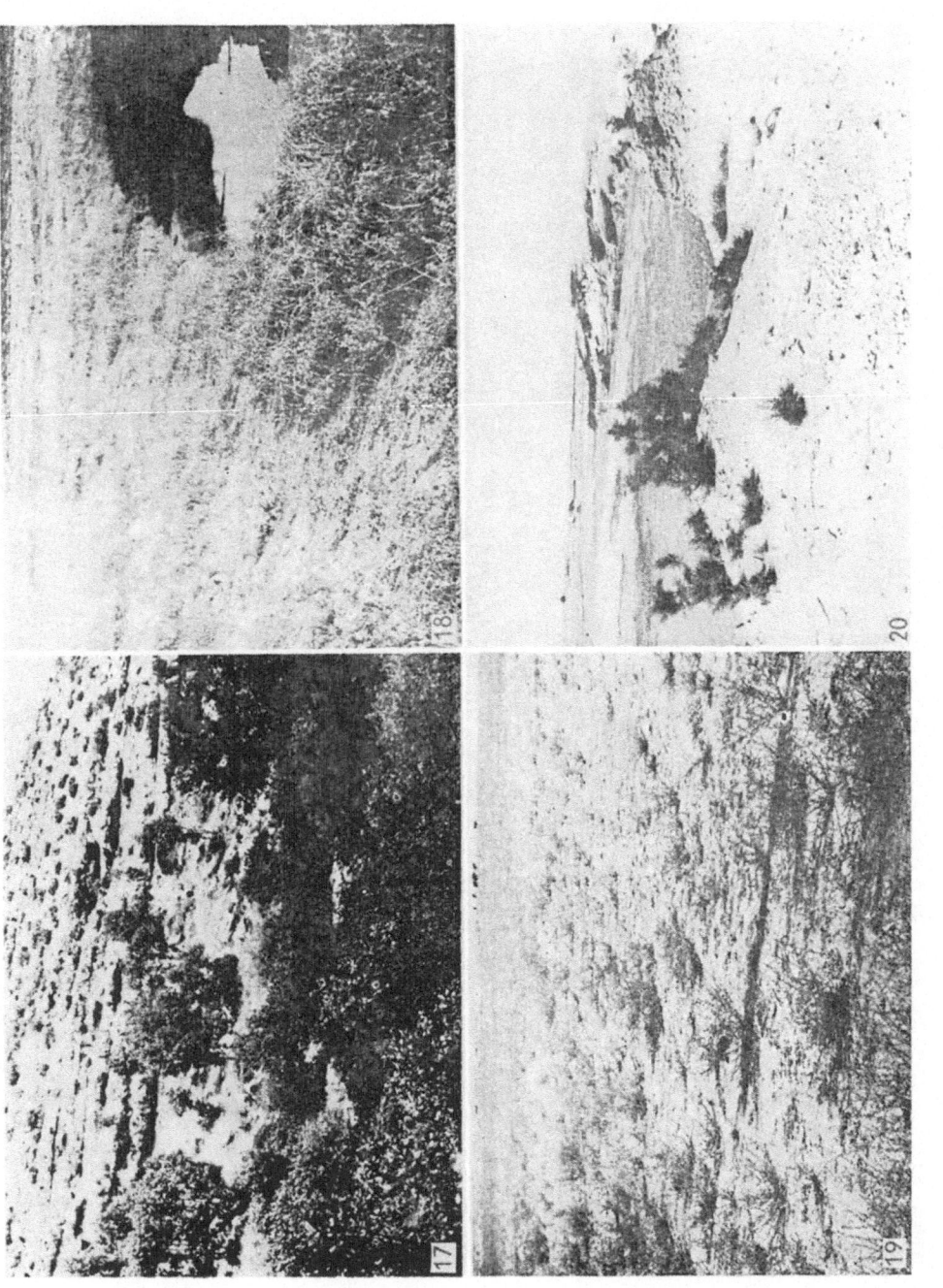

17. Remnants (cut) of oak woodland, mountains, Kurdistan, Iraq. — 18. Steppe, a river valley, submountains, Kurdistan, Iraq. — 19. Semidesert, Kurdistan, Iraq. — 20. Desert river valley, Lower Iraq.

21. *Chaitophorus* sp. on *Salix*, mummies, parasite: *Lysiphlebus salicaphis.* — 22. *Thelaxes suberi* on *Quercus* sp., mummies, parasites: *Lysiphlebus thelaxis* and *Trioxys quercicola.* — 23. *Aphis solanella* on *Solanum nigrum*, mummies, parasite: *Lysiphlebus ambiguus.* — 24. *Aphis fabae* on *Vicia faba*, mummies, parasite: *Lysiphlebus fabarum.* 25. *Aphis punicae* on *Punica granatum*, mummies, parasite: *Lysiphlebus ambiguus.*